THEISTIC EVOLUTION
The Teilhardian Heresy

BY THE SAME AUTHOR

Cosmos and Transcendence:
Breaking Through the Barrier of Scientistic Belief

Science and Myth:
With a Response to Stephen Hawking's The Grand Design

The Quantum Enigma: Finding the Hidden Key

The Wisdom of Ancient Cosmology:
Contemporary Science in Light of Tradition

Sagesse de la Cosmologie Ancienne

Christian Gnosis:
From Saint Paul to Meister Eckhart

De la Physique à la Science-Fiction

Wolfgang Smith

THEISTIC EVOLUTION

The Teilhardian Heresy

⊕

> "That henceforth we be no more children
> tossed to and fro, and carried about with
> every wind of doctrine..."
> —Ephesians 4:14

First published in the USA
© Wolfgang Smith 2012
Angelico Press / Sophia Perennis, 2012
All rights reserved

Series editor, James R. Wetmore

No part of this book may be reproduced or transmitted,
in any form or by any means, without permission.

For information, address:
Angelico Press / Sophia Perennis, 4619 Slayden Rd., NE
Tacoma, WA 98422
angelicopress.com
sophiaperennis.com

978-1-59731-133-5 pb
978-1-59731-134-2 cloth

Jacket Design: Michael Schrauzer
Cover image: *Creation (with Natural Phenomena)*, 1971
Courtesy of the artist, Cliff McReynolds

To the Blessed Virgin Mary
Mother of God
and Unfailing Help of Christians

My people have been a lost flock,
their shepherds have caused them to go astray,
and have made them wander in the mountains:
They have gone from mountain to hill,
they have forgotten their resting place.

—Jeremias 50:6

TABLE OF CONTENTS

Preface 1

Introduction 3

1. Evolution: A Closer Look 15
2. Forgotten Truths 38
3. Complexity/Consciousness: Law or Myth? 53
4. In Search of Creative Union 73
5. The Omega Hypothesis 94
6. The God of Evolution 115
7. Biblical Fall and Evolutionist Ascent 147
8. The New Eschaton in Historical Perspective 166
9. Socialization and Super-Organism 190
10. The New Religion 222

Appendix: The Riddle of Genesis 2:4–5 249

Pierre Teilhard de Chardin: Biographical Facts 253

Acknowledgments 255

List of Abbreviations: Writings of Teilhard de Chardin 256

Index of Names 257

PREFACE

THE PRESENT EDITION differs somewhat from the original, which appeared in 1988 under the title *Teilhardism and the New Religion*. It contains, first of all, a new Introduction which clarifies both the scientific and the theological status of theistic evolutionism, and introduces its founder, Pierre Teilhard de Chardin. The first chapter too has been updated and revised.

It should be noted that in the quarter century that has now elapsed since the book first appeared, the need for an exposition of this kind has significantly increased. When the tenet of theistic evolution is conveyed to the faithful by their clerical mentors as Catholic truth, it is high time to set the record straight! And if, for whatever reason, the ecclesiastic guardians of the Faith fail to do so, the task falls upon the laity: it is then *our* duty to defend the authentic Christian faith against the heresies of our time.

By way of a first orientation it can be said that what presently confuses and misleads the faithful above all are pseudo-philosophical notions masquerading in scientific garb. It is this spurious pretension to be "science-based" that renders these tenets virtually sacrosanct in the eyes of the populace, and explains why even theologians of rank have been misled. Does not our science work *"signs and wonders"* that could indeed *"deceive even the elect"* as Christ has foretold?

Although it is not just a single tenet but an entire syndrome of scientistic myths that presently befuddles the faithful, I surmise that evolutionism plays a central role in this collective process of subversion. In its theistic form, at any rate, it is doubtless the aberrant teaching which today most profoundly impacts Christianity. And this brings us back to Teilhard de Chardin, the architect and most ardent proponent of that hybrid dogma, who in the sixties came to be seen as a kind of modern-day prophet. It was he, moreover, who clearly perceived what most observers fail to see: the fact, namely, that a theistic evolutionism is irremediably incompatible with orthodox Christianity. For the French Jesuit it was however the latter—and not the heresy—that needs to go: the present stage of "human evolution," he thought, demands as much. And as he confided to a few of his most intimate friends, he perceived it to be his mission to usher in what he euphemistically termed "a new Christianity." We shall leave unconsidered the delicate question to what extent that objective has in fact been realized: it suffices us to defend the "old" Christianity which goes back, through the Fathers of the Church, to Christ Himself.

CAMARILLO,
FEAST OF THE
ARCHANGEL GABRIEL

INTRODUCTION

THE TENET of theistic evolution, as its very name implies, is the offspring of two disparate disciplines—science and theology, namely—and may consequently be viewed from either side. Let us begin on the side of science by recalling the gist of Darwin's famous theory. The basic idea is as simple as it is bold: one species, supposedly, can give birth to another by way of small random mutations. The stipulated scenario is obviously "slow" and hence of enormous duration, and requires a principle of selection to determine which mutations are incorporated into the line of descent. And here again the idea is simple: it is summed up in the phrase "survival of the fittest." Let me further recall that Darwin supported his theory in *The Origin of Species* by citing copious examples of species inhabiting the Galapagos Islands which had deviated from their mainland ancestors through the acquisition of characteristics needful to survival in the new environment.

More than a century and a half has now elapsed since this theory was first promulgated, and apparently it has taken about that length of time for the inherent difficulties to manifest. One must remember that, in Darwin's day, the science of biology was as yet in its infancy. Most notably, genetics—the biology of descent!—did not yet exist: the foundation was laid by Gregor Mendel in 1865, and his work remained unnoticed for some thirty five years. It then took another half century before the actual structures that carry the hereditary traits—the DNA contained in the nuclei of cells—came into view. And it turns out that the more biology we know, the more difficult it becomes for the Darwinist to stand his ground. The interested reader may consult the serious anti-Darwinist literature—which in recent decades has been growing by leaps and bounds—to learn exactly what the problems are that now threaten the theory, and in the opinion of many, have rendered it untenable.

For the purpose of this Introduction, three simple points will suffice. First, the evidence for the existence of evolutive transformations, as given in *The Origin of Species*, pertains to what has come to be called "microevolution," a kind which is severely limited in its scope. Whether microevolution does or does not transgress the bounds of a species depends of course upon the definition we assign to this elusive term. Yet it appears that the authentic species does prove to be inviolable. The Galapagos finches, notwithstanding their unusual plumage and beaks, are still finches; and as a matter of fact, no *bona fide* "transformation of species" has ever been observed. And this sharpens the debate: what henceforth stands at issue is *macroevolution*. And here the theory runs into two main problems, the first being what some have termed "fossil stasis." Not only are the intermediary forms demanded by Darwin's theory nowhere to be found, but it happens that the fossil record is characterized throughout by a pattern of stasis, interrupted here and there by the sudden emergence of new morphological forms. In a word, the paleontological evidence unquestionably repudiates the Darwinist theory: that is the first major problem, which has been recognized for quite a long time. The second came to light more recently: with the emergence of molecular biology, to be exact, a science which provides mathematically sharp examples of what Michael Behe terms "irreducible complexity." The idea is simple. "By *irreducibly complex*," writes Behe, "I mean a single system composed of several well-matched parts that contribute to the basic function, wherein the removal of any one of the parts causes the system to effectively cease function." What this means is that an irreducibly complex structure could not have evolved by way of chance mutations selectively assumed into the genetic line: on the molecular level the question submits to mathematical analysis through the calculation of probabilities, and thus at last renders possible a rigorous refutation of the Darwinist hypothesis. This line of inquiry is presently being pursued by advocates of what has come to be known as "intelligent design." More cogently perhaps than any other discipline, ID research demonstrates the impossibility of the evolutionist claim.

Why, then, in the face of mounting counter-evidence, is the Darwinian theory not only retained, but pronounced in high quarters to be a scientifically sacrosanct truth? The answer to this puzzling question has been given with the utmost clarity by a leading evolutionist himself. I am referring to Richard Lewontin, who avers that it is "our *a priori* commitment to material causes" that drives the Darwinist: "and that commitment," he tells us, "is absolute, for we cannot allow a Divine Foot in the door." Which brings us finally to our proper subject, namely "theistic evolution": a new kind of Darwinism, which not only does allow "a Divine Foot in the door," but maintains that "a Divine Foot" is in fact needed if evolution is to take place. Instead of replacing God as Creator by the evolutive process, theistic evolution affirms that this process is actually the means by which God does create: "God creates by evolution," so the dictum goes. Strange as it may sound, Darwin's atheistic theory has thus metamorphosed into a theistic doctrine, espoused today by major segments within the Christian world. The fact is that theistic evolutionism, in conjunction with the so-called big bang scenario, is nowadays taught in seminaries and widely disseminated to the faithful as the enlightened up-to-date cosmogony, which in effect replaces what is waved aside as the "literal sense of Genesis."

We shall come to the theological issues presently. But first it behooves us to reflect somewhat upon the propriety of theistic evolutionism as such: of bringing God into the picture precisely as a kind of *deus ex machina*, missioned to make Darwinian evolution work. Instead of letting the Darwinist hypothesis fail on scientific grounds, it seeks to bolster that now faltering theory by the *ad hoc* postulate of divine intervention, for which, to put it mildly, there is not a shred of theological rationale. In a word: theistic evolutionism compounds bad science with spurious theology. One fails, moreover, to recognize that once God has been affirmed, there is no further need and no reason on earth to maintain the transformist hypothesis: the unfounded notion that one species "evolves" into another. If the *raison d'être* of that far-fetched idea is indeed to proscribe "a Divine Foot in the door," is it not the height of folly, on the part of Christian apologists, to bolster that atheistic and now discredited hypothesis through the no less gratuitous postulate that God Himself steps in to consummate the anti-God scenario? One is hard-pressed to name a second doctrine as flagrantly inane! Worst of all, however, it turns out that this ill-conceived tenet, promulgated chiefly by men of the cloth, comes at a terrible price, for it is tantamount, in the final count, to *a denial of the Christian truth*: in a word, theistic evolutionism is in fact *heresy*. This is what I now propose to show.

One looks askance, these days, at the so-called "literal sense" of Genesis, and in particular of its first three chapters, as if that Biblical tale were simply a myth: the fantasy, that is, of a primitive age. One forgets that if, indeed, there *was* a Creation, and if Adam and Eve *did* initially reside in Paradise, then that part of the story, insofar as it transgresses the categories of the post-Edenic realm that constitutes *our* world, *must* evidently be in a sense "mythical." In expounding these primordial themes, the Biblical author was consequently forced to employ language in a manner all its own: to speak of Adam is not, after all, like speaking of a long-deceased grandfather! Even the most simple-minded interpretation, therefore, of the Biblical text, so long as it comprehends in some measure what it is meant to convey, *cannot* be "literal" in the ordinary sense. And if, indeed, we are so earth-bound, and so utterly prosaic, as to read the Good Book as if it spoke simply of water and earth, of trees and apples and serpents, we do so to no avail: we simply miss the point. And let me note, in passing, that a good deal of "Biblical criticism" these days springs from nothing more sublime than an inability to understand the text as it is meant to be understood. One basically takes the world-view of contemporary science as his reference frame, determined from the outset that the message must fit these confines or be truncated till it does. And to be sure, what remains after the text has been thus "demythologized" is bereft of all ontological significance, and can at best be viewed as a kind of moral teaching adapted to untutored minds. But let us go on: my point is that the most profoundly definitive texts in both Testaments—beginning, certainly, with the first three chapters of Genesis—are meant precisely *not* to fit into *any* post-Edenic reference frame, be it indigenous to the Stone Age or to the 21st century. One might say that the very purpose of Holy Writ is to bring us *out*

of what the poet terms "this narrow world," which remains such, metaphysically speaking, despite the quantitative immensities proffered by our astrophysical cosmologies. And that incomparable expansion of outlook—that veritable *liberation!*—is indeed what the much-despised "pre-critical" or so-called "literal" reading of Genesis accomplishes for those who "have ears to hear." The first three chapters, in particular, teach us that *this* world came into existence as the result of a *Fall*, and thus bring into indirect view a transcendent realm, a world beyond *this* world, which exceeds and in a way encompasses the latter, and in so doing dwarfs all our accustomed immensities. It is this fundamental breakthrough, precisely, that enables us to contemplate—"*as through a glass, darkly*"—the central truths of our faith, beginning with the existence of God and that Kingdom said to be "*not of this world.*" So far from being "mythical" in the pejorative sense, or inessential, the opening chapters of Genesis constitute the very basis for the exposition of Christian truth: this is actually where the Christ-story begins. Remember: Christ is the Second Adam, who came into this fallen world to redeem mankind from the Sin of the First. Let us understand it well: Redemption presupposes the Fall! And it matters not a whit whether Darwin, Einstein or Hawking concur: *this is what Christianity teaches, and what Christians believe.*

We need now to ask ourselves whether the tenet of theistic evolution is or is not compatible with the sacred truths of which the first three Genesis chapters tell. Is it conceivable, in other words, that *both* accounts of man's origin could in fact be true? But no sooner is the question posed than the answer stares us in the face: if man, body and soul, originated in a *transcendent* state—a state which eludes the bounds of *our* world or universe—then most certainly he did *not* originate so many million years ago on what science identifies as a planet in our galaxy. In other words, to accept the latter position is to deny the aforesaid transcendence as affirmed by the sacred text. So too consider the episode of the Fall: God's commandment not to eat the fruit of a certain tree "*lest ye die*," followed by Adam's transgression and expulsion from Paradise. What conceivable sense does any of this make from an evolutionist point of view? How does a theistic evolutionist, in particular, interpret "*the *tree in the midst of the garden*," by means of which *death* entered the world? What the Fathers perceived as a divinely inspired teaching of incomparable profundity, the evolutionist is evidently compelled to dismiss as a mere fable: a didactic tale of some kind, presumably, addressed to a childlike people of a primitive age. What an impoverishment! And what presumption this, what shameful overweening! Most flagrant of all, the evolutionist interpretation—be it ever so "theistic"—misses the most crucial point: that *death*, namely, is the consequence of *sin*, the breach of a divine commandment: "*thou shalt not eat of it; for in the day that thou eatest thereof, thou shalt surely die*." Is this too a mere fable? Evidently two thousand years of Christian tradition declare that it is not. As a matter of theological fact, the nexus between "sin" and "death" proves to be no less fundamental in the New Testament than in the Old: Redemption, most certainly, precedes Resurrection. And let us not fail to note that this Cycle or History, marked by Fall, Redemption and Resurrection, involves not only mankind, but affects the creation or universe at large: for strange, if not impossible, as it may strike our science-conditioned mind, Man stands at the absolute Center—no Copernicus, no Einsteinian relativity here!—and by virtue of his unique function and relation to God affects the creation at large. Think of it, then: if indeed "*death came by man*" as St. Paul testifies, how could man have originated by way of an evolutionary chain entailing death upon death over millions of years?

We need not belabor the point: the Biblical and the Darwinist account of man's origin, one sees, are as different as night and day; and this holds true even if the latter is amended by the incongruous stipulation that God Himself lends a hand, as it were, in the evolutive process. It matters not whether the theological advocates of theistic evolutionism bring into play the notion of a "Babylonian creation myth" to render the Genesis tale innocuous or employ some erudite stratagem to defend their position theologically, the fact remains that the tradition going back through the Fathers to Christ Himself has been compromised, the foundational teaching rejected, and the dogma of Scriptural inerrancy denied. And this, to be sure, is heresy, pure and simple, whether the theistic evolutionist knows it or not.

But in fact he doesn't: to the "liberated" theologian every tenet of orthodoxy has become fair game. The heresy now, it seems, is to believe that there *is* such a thing! Beyond a certain point of "liberation," just about anything goes. The objective criteria, which for some two thousand years had defined and protected theological orthodoxy, have then given way to the subjective norms of contemporary pundits, the so-called *periti* or "theological experts" of our day; and this accounts evidently for all kinds ecclesiastic novelties and "up-to-date" teachings, dispensed nowadays from pulpits across the land. For the Christian observer, on the other hand, who happens *not* to be thus "liberated," all this newness is hardly a cause for rejoicing. He perceives this trend, rather, as a breach of God-given norms which forthwith plunges the perpetrators into a man-made fantasy world. One may wonder whether that kind of "theology" will not inevitably end either as a mere sponsor of human friendliness and social service or as a psychotherapy. Authentic theology, on the other hand, is neither, or better said, is incomparably more. And what prevents authentic theology from sliding into the "human-all-too-human" are in fact the Bible- and Tradition-based criteria of orthodoxy. Let us understand it well: to breach these criteria is *heresy*.

This brings us back to theistic evolutionism, the status of which can now be clearly recognized. As once, according to St. Jerome, "the Church groaned to find itself Arian," so apparently it "groans" today to find itself "evolutionist."

Who, then, is the new Arius, the grand architect of this heresy: the genius who inspired millions with the idea of theistic evolution, beginning with duly emancipated circles within the traditionally cautious and conservative Roman Catholic Church? As one knows very well, that person was none other than Pierre Teilhard de Chardin, the Jesuit paleontologist who for many years was silenced by Rome, but all the while continued to write and make converts among ecclesiastic colleagues of a similarly "progressive" bent. And when at last his teaching emerged into the light of day—around the time of the Second Vatican Council—it was enthusiastically received within the Catholic world, beginning with the erudite, the intellectual leaders. First to be affected was the Jesuit Order; yet soon enough the Teilhardian message was disseminated by way of lectures and seminars at Catholic institutions of learning throughout Europe and America. From thence it spread to ever widening circles, and everywhere it aroused a kind of intellectual excitement, as if some deep chords had been struck. Clearly, the moment was opportune: these were, after all, the heady days following Vatican II. The stage had now been set, and a goodly portion of the believing and unbelieving world was ready and most willing to imbibe the message: to partake freely of the new wine. Translated ere long into twenty-seven languages, the posthumous treatises of the once silenced Jesuit came thus to exert an inestimable influence. In the end millions were affected in varying degrees.

Today, of course, the rage has abated. By now Teilhard's ideas—which in the sixties had seemed so revolutionary—have lost much of their capacity to shock and to enthrall. And whereas the actual logic and elaborate terminology of his theory have been roundly forgotten, its gist has become almost commonplace. It is expressed in the dictum *"God creates through evolution"*: that is the *mantra* which has now gained currency throughout the Christian world as an article of belief that profoundly affects the way millions understand and practice their faith.

But then, why interest oneself in an analysis of a now forgotten theory? If the aforesaid *mantra* is basically all that survives, why bother with the rest? Apart from the fact that this forgetting is by no means absolute, and that a certain residue of the Teilhardian doctrine remains active in the collective memory, affecting the thought of contemporary theologians, the principal reason, I say, is that Teilhard de Chardin, more keenly perhaps than anyone else, perceived the implications for Christianity of what he termed "the truth of evolution." Where others have been quick to graft that presumed "truth" onto a duly truncated theology, Teilhard recognized, with singular clarity, that in order to merge the two doctrines, one needs to reformulate Christian theology from the ground up: it is a question of "laying new foundations to which the Old Church is to be gradually moved" to put it in his own revealing and fateful words. But whereas, for an evolutionist, this laying of "new foundations" constitutes a glorious advance—the replacement of a primitive mentality that has outlived its use—for an orthodox Christian such a deviation constitutes of course outright heresy. Paradoxical, thus, as it may seem, the very argument which led Teilhard to elaborate a theology of theistic evolution brings to light the crucial fact that a doctrine of that kind is perforce heretical.

Yet, undeterred by that decisive recognition, Teilhard evidently perceived this "laying of new foundations" as the very mission of his life. And as we shall discover in the final chapter of this book, that passion, which came to dominate him, appears to be grounded in a preternatural experience of some sort, which years later, when he was writing *The Heart of Matter*, caused him to exclaim: "How is it, then, that as I look around me, still dazzled by what I have seen, I find that I am almost the only person of my kind, the only one to have *seen!*" On fire with an idea, Teilhard was evidently bent upon setting the rest of humanity ablaze as well. And whereas this universal conflagration has evidently not materialized, the impact upon mankind of that "one man who *sees*" has been nonetheless stupendous.

What exactly, then, was Teilhard de Chardin? Authentic prophet, to be sure, he was not. A genius, then, or perhaps a brilliant charlatan? Actually, he was both. One needs namely to distinguish between his core recognitions, underpinning what he perceived to be the need for new theological foundations, and his scientist fantasies, decked out in what Peter Medawar of Nobel laureate fame refers to as "that tipsy, euphoric prose-poetry." Whatever else one may say concerning such bizarre conceptions as Teilhard's "radial energy," or his so-called "law of complexity/consciousness," the fact remains that these scientific-sounding fictions sufficed to convince a multitude of theologians that his was indeed a science-based doctrine. We shall presently see, chapter by chapter, that the theologians have been misled. Yet it is doubtless this presumption of "scientific evidence for theistic evolution" that enabled the doctrine to attain dominion within the educated Christian world.

What above all concerns us, however, is the inherent contradiction between a Darwinist evolutionism, however theistic, and orthodox Christianity. It is one thing to reject the tenet of theistic evolution, and quite another to understand in depth why its theological implications across the board are radically unorthodox; and no one, it seems to me, has shed more light on that vital question than Teilhard de Chardin. He did so, of course, not to protect Christian orthodoxy, but for the opposite reason: to prove, namely, that tradition-based orthodoxy must go. But that is another question: what concerns us is the fact that in these profoundly anti-Christian reflections we encounter—not a pretender spouting "tipsy, euphoric prose-poetry"—but an intelligence of high rank: not the charlatan, but indeed the genius side of the man. And let me add that these tendentious dissertations bring to light not only the theological implications of theistic evolutionism, but, just as importantly, its effect upon the spiritual life. Teilhard gives us to understand that what is traditionally regarded as the following of Christ has now become obsolete; for the enlightened Christian it is not a question of *following* but literally of *creating* Christ: creating Him by way of *evolution*. Whereas Christ has always been conceived as both Alpha *and* Omega, He is now reduced to Omega: this half-truth is all that is left.

No wonder Teilhard disapproves of Christian piety as understood and practiced up till now; even the Beatitudes, as we shall see, do not escape censure at the hands of this strange priest! Basically what remains of the spiritual life is communal action, the kind that promotes "socialization," to use one of Teilhard's peculiar terms; after all, what counts, from an evolutionist point of view, is not the individual, but the species. I would add that these Teilhardian dissertations, based squarely upon his theistic evolutionism, can teach us much regarding the "liberated" religious life in general, what some perceive to be new forms of spirituality in keeping with the times. It should come as no surprise, moreover, that Teilhard is well disposed towards the disciples of Karl Marx, and in fact considers them fellow travelers, whether they know it as yet or not. In short, one finds in the Teilhardian opus a preview, as it were, of tendencies and aberrations affecting the spiritual life of the laity as well as of religious orders which have manifested following Vatican II, or are still in process of manifesting. Overall Teilhard de Chardin belongs clearly to that class of profound visionary writers who can help us to understand the world in which we live.

Finally I wish to impress upon the reader—as I have on previous occasions—that what I place in his hands is not an academic exercise, but is meant to serve an eminently practical end. As Christians we ought never to forget that we are here on earth for no other purpose, finally, than to save our soul; and in company with the Fathers of the Church, I believe that to this end *doctrinal orthodoxy* is imperative. Yet, whether we know it or not, that orthodoxy stands today under attack, from without and from within, as perhaps never before. Granting, moreover, that we are currently plagued by more than a single heresy, I surmise that theistic evolutionism, in particular, plays a pivotal role inasmuch as it contradicts the Biblical revelation more directly and profoundly than the rest. I therefore offer this treatise on the teachings of Pierre Teilhard de Chardin to my fellow Christians as a kind of homeopathic remedy, a medicine distilled, if you will, from the very teachings that have brought on the disease. But metaphors aside, what I wish to convey is a definitive theological *and* scientific refutation of the Teilhardian doctrine: that curious science-fiction theology which has bedeviled so many and exerted a devastating influence upon the Church.

What is called for, above all, is an exposition of Christian truth that vanquishes the heresy. The problematic of the Teilhardian teaching serves thus as a point of departure for the application of universal and sovereign principles, at once metaphysical and theological, to fundamental issues, such as the nature of time and eternity, the twofold distinction between spirit, mind, and matter, or the meaning of history in light of Revelation. In a way the present monograph follows the lead of the ancient *"adversus haereses"* treatises, wherein polemic argument and didactic exposition intertwine. From start to finish I have been motivated and spurred by one paramount concern: namely, to safeguard and render intelligible a living legacy which is incomparably more than a mere philosophical system or formalized theology. And it happens that the Teilhardian doctrine lends itself splendidly to this end: the greater the heresy, the more profound its service in the quest of truth! Did not Arius of old contribute more perhaps than any other heresiarch to the unfolding of Christology? The sovereign truths of Christianity, of course, have been rendered manifest ages ago; our task today is but to recall and contemplate these God-given truths—and love them with all our heart.

1

EVOLUTION: A CLOSER LOOK

WE HAVE ALREADY taken a first glance at Darwin's theory, and have noted that all is not well. As the reader will recall, that summary account began where it must: with the distinction, namely, between macro- and microevolution. No one questions the reality of microevolutionary transformations: the debate, then, is about macroevolution. And one might add that more, perhaps, than anything else, it was the failure to distinguish clearly between these two disparate conceptions that obfuscated the issue during the period immediately following the promulgation of Darwin's theory. We should also note that the latter half of the nineteenth century was intellectually dominated by a victorious physics, which however proved in the end to be utterly naïve. One knows today that it was an uncorrected physics of continuity that allowed facile extrapolation, say from nanometers to light-years: is it any wonder, then, that biologists, too, tended to be overconfident?

In any case, not until the basic distinction between micro- and macroevolution has been put in place is it possible to make cogent pronouncements on the subject of Darwinism. And the salient fact is this: *if by "evolution" we mean macroevolution* (as we henceforth shall), *then it can be said with the utmost rigor that the Darwinist claim is bereft of scientific sanction.* Now, to be sure, given the multitude of extravagant claims regarding evolution promulgated by leading authorities with an air of scientific infallibility, this may indeed sound strange. And yet the fact remains that there exists to this day not a shred of *bona fide* scientific evidence in support of the thesis that macroevolutionary transformations have ever occurred.

One is reminded in this connection of Colin Patterson, a senior paleontologist at the British Museum of Natural History, who in recent times has shocked the public—and the experts—by challenging anyone to come up with even a single authentic example of a macroevolutionary transformation. Despite a voluminous amount of mail response, most of it irate, it seems that no one was able to come up with one. When a reporter questioned Patterson on this point, he was told that one person had written in to say that the polar bear is descendent from the brown bear. "Do we really know this?" the reporter asked. "No" replied the paleontologist.

Not only is there no conclusive evidence of any kind that a macro-evolutionary transformation has ever taken place, but it happens that the more biology one knows, the more certain it becomes that no such transformation *can* occur. The two major difficulties, as we have noted, are fossil stasis and irreducible complexity. But there are a host of other problems, which not only can, but have indeed "filled books."[1] How, then, do the advocates of Darwinism defend their position in the face of this mounting counter-evidence? The basic strategy, strange to say, is to postulate additional hypotheses. These so-called *ad hoc* claims are introduced either to explain away conflicting facts, or to render neutral facts corroborative. To give an example of the former, let me recall that the absence of "intermediate forms" in the fossil record—which, according to Darwin himself, would suffice to disprove his theory[2]—was countered by postulates of cryptogenesis or "hidden evolution," one of which, as a matter of fact, was proposed by Teilhard de Chardin himself. As Phillip Johnson, the Berkeley law professor and author of *Darwin on Trial*, pointed out: "Darwinism apparently passed the fossil test, but only because it was not allowed to fail."

---

1. To mention at least a few titles from the serious anti-Darwinist literature: Michael J. Behe, *Darwin's Black Box: The Biochemical Challenge to Evolution* (New York, NY: Free Press, 1996); Michael Denton, *A Theory in Crisis* (Bethesda, MD: Adler & Adler, 1986); Douglas Dewar, *The Transformist Illusion* (San Rafael, CA: Sophia Perennis, 2005); Phillip E. Johnson, *Darwin on Trial* (Downer's Grove, Ill.: Inter-Varsity Press, 1993); Richard Milton, *Shattering the Myths of Darwinism* (Rochester, VT: Park Street Press, 1997).

2. *The Origin of Species* (Chicago: Britannica, 1952), pp. 163–64.

As to the second category of *ad hoc* hypotheses—the kind that turns neutral facts into "evidence for evolution"—let me recall the so-called "vestigial organs." In Darwin's day much was made of these imagined "remnants" of an earlier evolutionary phase, which could supposedly be identified as such by what Darwin termed "the plain stamp of inutility." Just about any anatomical structure the function of which was insufficiently understood—from the pineal glad to the vermiform appendix— could thus be held up as a degenerate and no longer functional vestige of an ancestral form, and paraded by Darwinists as evidence that evolution has occurred. Meanwhile it turns out that the list of these "vestigial organs" becomes shorter year by year: in light of increased knowledge "the plain stamp of inutility" exhibits a tendency to disappear. One should also bear in mind that what Darwin's theory actually demands are not vestigial but "nascent" organs, such as eyes that do not yet see and wings that do not yet enable flight. But it appears that no living species exhibits a structure of that kind: not a single such "organ to be" has ever been identified.

The centerpiece of Darwinist theory must evidently be a genealogical tree of some kind, purporting to lay out the main lines of evolution, such as the celebrated specimen proposed by Ernst Haeckel in the early days. But whereas the leaves of such a tree represent actual species, the branches and trunk prove in the end to be fictitious. Haeckel's so-called "protovertebrates, protoselachians, protoamniata and protomammals," for example, have left behind not so much as a single fossil to document that there ever was such a creature; as someone has put it: these Darwinists were not embarrassed "to populate the ancient seas and continents with schemata." Worse still, it turns out that the game can be played in many different ways, depending upon the particular structures one pursues in this imagined evolutionary ascent. In the end, everyone can just about pick his own tree!

Along with his genealogical tree, Haeckel has left us a famous principle known as the biogenetic law, which proclaims that the embryo in its ontogeny recapitulates the evolutionary stages through which the species itself has passed. Let me recall that this too was enthusiastically received and regarded for many decades as an established "law of nature"—until it became clear that such things as the so-called "fish heart" and "gills" are formed in the mammalian embryo precisely to ensure an adequate blood supply. Generally speaking, "recapitulated" structures exist for basically the same reason as does a scaffold in the building of a house: for no more esoteric cause, namely, than the fact that they are needed! Following this sobering recognition, the famed "law" was silently discarded, though it survives to this day in some niches of the academic world. In the end that stipulated "ascent of the evolutionist tree" on the part of the embryo has proved to be no less imaginary than the tree itself.

Meanwhile it turns out, in light of findings pertaining to molecular biology, that the very idea of a "genealogical tree" is misconceived: let me explain. With the discovery of DNA, in 1953, it has become possible to speak of "distance" between living species in a precise quantitative way: for example, in terms of so-called percentile difference sequences associated with certain proteins, such as cytochrome C; and this has led, first of all, to the rediscovery of a fundamental fact, which in a way goes back to Aristotle, and has always been recognized by taxonomists: namely, that the biosphere breaks up into well-defined and widely separated classes, organized according to a hierarchic (as opposed to a sequential) principle. It happens that on the molecular level these separations, and this hierarchic order, stand out with mathematical precision. Consider, for instance, that a carp is 13 percentile sequence differences (based upon cytochrome C) removed from a horse, 13 from a turtle, and 13 from a bullfrog. Now, the fact that a fish is thus "equidistant" from a mammal, a reptile and an amphibian is hardly compatible with the evolutionist postulate that mammals have descended from reptiles, reptiles from amphibians, and amphibians from fish. In light of such recognitions it becomes apparent that a "tree model" fits the taxonomic facts about as well as a "chain model" suits a lattice, which is to say, of course, that it fits not at all. Indeed, as Michael Denton points out:

> The only way to save evolution in the face of these discoveries is to make the ad hoc assumption that the degree of biochemi-

cal isolation of the major groups was far less in the past....
There is, however, absolutely no objective evidence that this
assumption is correct.[3]

What above all, however, militates against the Darwinist hypothesis is the phenomenon of "irreducible complexity" on a molecular scale. As noted earlier, it is the stupendous intricacy of molecular structure and function, encountered in even the simplest forms of life that permits one to quantize the "astronomical unlikelihood" of the Darwinist scenario. Beyond a certain point of "smallness"—which of course is not uniquely definable—a probability "close to zero" tells us that the event in question is physically impossible: if that were not so, nothing on earth could in fact be ruled out. To give at least one example, let us consider the so-called bacterial flagellum, a justly famous instance of irreducible complexity, which is all the more remarkable inasmuch as it pertains to one of the simplest forms of life. Here are the facts: the flagellum is a molecular device whose function it is to propel the bacterium through a liquid ambience in its quest for nutrients. The flagellum itself constitutes a kind of molecular rotary paddle or propeller which is relatively simple; what proves to be complex is the molecular device that renders it functional. This feat of nanotechnology comprises an acid-powered rotary engine, endowed with what amounts to a rotor, a stator, a drive shaft, bushings and O-rings. To accomplish its function in the face of Brownian motion the device must attain angular velocities close to 10,000 rpm, and be able to reverse direction in less than one hundredth of a second. In addition to its acidic fuel and structures to store and distribute that fuel, its operation requires auxiliary devices for detection and control. In all, about two hundred forty different proteins are needed to make a functional flagellum.

What is one to say? Surely Darwin himself, had he known even the half of it, would never have proposed his famous idea! At long last, moreover, mathematics comes into play: it is possible finally to do more than wave one's hands. One is now able to calculate upper bounds for the probability of a Darwinist origin; and as one might

---

3. *Evolution: A Theory in Crisis* (Bethesda, MD: Adler & Adler, 1986), p. 29.

expect, in the case of structures comparable to the bacterial flagellum, that upper bound amounts to outright impossibility; as someone has suggestively put it, one might as well suppose that a cyclone, passing through a junk yard, gives rise to an automobile! Now, faced by irreducible complexities of such magnitudes in even the most rudimentary forms of life, how can anyone continue firmly to believe in Darwinism? Since it is evidently impossible to do so on scientific grounds, one is forced to conclude, with Richard Lewontin, that what drives the contemporary Darwinist is indeed "our commitment to material causes," which in fact "*is absolute.*" What confronts us in the Darwinism of our day is thus no longer science, properly so called, but proves to be, ultimately, a kind of religion: a counter-religion, to be exact.

Let us then consider what authentic theology has to say regarding the origin of life on Earth. The central and most crucial point has been articulated by St. Augustine: "Beyond a doubt the world was not made *in* time, but *with* time."[4] And this in itself settles the matter: God does *not* create by means of evolution, for the simple reason that evolution takes place *in* time. Now, a theistic evolutionist might object to this conclusion on the grounds that one needs to distinguish between the world as such and living creatures existing in the world: that whereas the world itself may indeed have been created *with* time, its creatures came to be *in* time nonetheless. But apart from the fact that the discourse from which this Augustinian dictum has been extracted rules out that interpretation, Scripture itself settles the issue once and for all: "*Qui vivit in aeternum creavit omnia simul*" ("He that liveth in eternity created all things at once").[5] As Meister Eckhart explains: "God creates the world *and all things* in this present now": in the *nunc stans*, namely, the "now that stands," which coincides with eternity. The very idea of "theistic evolution"—as expressed in the Teilhardian dictum "God creates

---

4. *De Civitate Dei*, 11:6.
5. Ecclesiasticus 18:1.

through evolution"—proves thus to be untenable metaphysically and indeed heretical to boot. Let the theistic evolutionist realize, once and for all, that God does *not* create *in time*. The creature comes first: *being* has precedence over time. In the memorable words of St. Augustine: "Let them see that without the creature there cannot be time, and leave off talking nonsense!"[6] So far from "evolving" over millions of years, creatures actually precede time itself. That precedence cannot, of course, be temporal: it is not a duration that can be measured by clocks. The point, rather, is that the creature, *in its being as such*, is not subject to time: for strange as it may seem, time itself derives ultimately from the creature. To quote St. Augustine once more: "God, therefore, in His unchangeable eternity, created simultaneously all things *whence times were to flow*."[7] Let me say it once more: the creature as such comes first, it *precedes* time.

Yet, to be sure, created beings come to birth in time: their entry into this world is localized in time and place. Each creature, in its cosmic manifestation, is thus associated with its own spatio-temporal locus, which is to say that it fits somewhere into the universal network of secondary causes. But yet it is not *created* by these causes, as the Darwinists think, nor is its being confined to that spatio-temporal locus: for its roots extend beyond the cosmos into the timeless instant of the creative Act. Let there be no doubt about it: the creature is more—incomparably more!—than its manifestation in space and time. It does *not* coincide with the phenomenon, the organism as conceived by the scientist, as one is inclined to suppose. Even the tiniest plant that blooms for a fortnight and then is seen no more is vaster in its metaphysical roots than the entire cosmos in its visible form: for these roots extend into eternity. And how much more does this hold true in the case of man! "*Before I formed thee in the womb, I knew thee.*"[8]

Admittedly, the *omnia simul* doctrine of creation is difficult to grasp. In a way it exceeds what man in his present state can actually

---

6. *Confessiones*, 11:40.
7. *De Genesi ad Litteram*, 8:39.
8. Jeremiah 1:5.

comprehend: for that doctrine corresponds, as it were, to the "perspective" of God Himself. There is however a second way of conceiving the origin of creatures, humbler but easier to comprehend; and that is the teaching of the *hexaemeron* or "six days" of creation, as related in the first chapter of *Genesis*. The perspective here, though intra-cosmic, is yet primordial or Edenic, as opposed to being localized in "peripheral" space and time, let alone in spacetime as conceived by the physicist. Yet it excludes the Center comprising the *omnia simul* of the creative Act, as indeed every intra-cosmic perspective must. Strictly speaking, the *hexaemeron* account refers thus to the effects of the creative Act, as distinguished from the Act itself. It refers to these effects, however, as perceived—not, to be sure, by the man of our day—but by the *anthropos* as such: not in fact by this-man-so-and-so, but by what may be termed Adamic or primordial man. And "relative" though it be—bound as it is to a creaturely perspective—that account of the creation is true: for what the *anthropos* as such perceives is *ipso facto* real. One could say that it is on that account "more" real—"more" true, if you will—than what *we* perceive. Yet, even so, it could be argued that *Genesis* itself refers also to the *omnia simul*, which is to say that it alludes at least once to the unbroken unity of the Creation which precedes the spatio-temporal divisions of our world. It does so in chapter 2, verses 4 and 5, which are doubtless among the most challenging to be found in the Old Testament. I venture to offer an exegesis, which has been relegated to an Appendix.

Let us then get back to St. Augustine's explication of the *omnia simul*: "God, therefore, in His unchangeable eternity, created simultaneously all things whence times were to flow." What confronts us here is a doctrinal statement of first rank, enunciated by one of the greatest Doctors of the Church. It happens, however, that it is spoken in a language which is no longer understood: our universe of discourse—and so, in a sense, our world—has actually shrunk. An entire dimension has in effect disappeared: the "vertical" dimension, namely, which enables us to speak of things "above" this universe, beyond this world perceived with our eyes and detected by means of scientific instruments. All authentic metaphysics and theology hinge evidently upon that dimension: the "above" is in fact their proper domain. For the past four or five centuries, however, leading philosophers and scientists have labored to dismiss that "above" as the fantasy of a pre-scientific mind; and their efforts have borne fruit: collectively and on average, at least, we have become denizens of the flatland, persuaded that this planar expanse is all. Two options then remain: one can dismiss the ancient doctrines outright, or reinterpret them in horizontal terms. For theology this means that the choice lies between atheism and heresy: there *is* no third option. And let us note that the latter poses by far the greater threat to religion: for whereas atheism denies from without, heresy corrupts from within.

Getting back to the subject of evolution, it has now become apparent, first of all, that the Darwinist is seeking "the origin of species" where in truth it cannot be found: for inasmuch as that origin *precedes* time, it transcends the existential plane to which his vision is confined. The theistic evolutionist, on the other hand, not only repeats that same Darwinist error, but contributes a fallacy of his own, more deadly than the first. Having identified the origin of species with events postulated to have taken place at different geologic times, he goes on to enunciate the credo of theistic evolutionism: "*God creates through evolution.*" This means that every hypothetical "leap" resulting supposedly in the emergence of a new species counts as a creative act of God. Gone is the scriptural *omnia simul*: the recognition that God does not create in temporal sequence, but "all at once"; and gone, too, the idea of ontological precedence: of created beings "whence times were to flow," to put it in St. Augustine's words. All that is left on the side of creation are the things subject to the flux of time, which in truth are not "things" at all; for as Plato observed long ago: "How can that which is never the same *be* anything?"[9]

For the theistic evolutionist, on the other hand, determined as he is to reinterpret the basic conceptions of theology—to cut them down to size, one might say—these problems do not exist. When even "trans-substantiation" can be serenely replaced by "trans-signification," as some duly liberated theologians have proposed, where

9. *Cratylus*, 439E.

does one draw the line? It appears that just about every concept of theology, even the most sacrosanct, can nowadays be "flattened," and so deftly in fact, that to the extent that the discrepancy is perceived at all, it is seen as an *aggiornamento*, the updating of an antiquated theology: entire journals have already been filled with the fruits of these labors. Theistic evolutionism, then, is a case in point: it answers to the prevailing trend. No one, therefore, need be surprised that nowadays the Teilhardian *mantra* rings true, that it is in fact music to contemporary ears. Exceptions are few: it is never easy, after all, to go against the Zeitgeist and its guardians, who know very well how to enforce their will.

Yet whatever the proponents of *aggiornamento* may ordain, the fact remains that authentic theology hinges upon the conception of ontological domains "above" the corporeal, which is to say that the physical or corporeal universe as such can finally be no more than the outer shell, as it were, of the integral cosmos. A figure of speech, this "outer shell," some will charge! Not exactly: a reference, rather, to a perennial icon which enables us to "visualize" that integral cosmos, perceive it "as through a glass, darkly." Now, that icon consists of a circle, the center, radii and circumference of which convey a metaphysical sense. Think of the first element, the center, as representing eternity: it is here, beyond the cosmic categories of space and time, that the "first birth" of every creature—its actual creation—takes place. From thence a kind of "descent along the radii" ensues, culminating in "the second birth": its entry into a "predestined" locus within cosmic space and time. The "movement," then, which precedes that second birth, is in a sense centrifugal or "downward," if one may put it so; and this constitutes in fact the true "evolution," which is not the creation of something that did not previously exist, but the manifestation in space and time of a created being. The very word, derived as it is from *evolvere*, designates evidently, not a making, but an *unfolding* of some kind. And let me point out that this authentic evolution is clearly recognized and definitively enunciated in the Patristic literature. St. Augustine, for example, speaks of the world as "pregnant with the causes of things unborn," to the end that the things in question may "break out and be outwardly created in some way by the *unfolding* of their proper measures."[10] The true evolution of creatures is thus precisely an "unfolding of their proper measures" resulting in their manifestation at a particular time and place. And one might add that this manifestation comes about, not by chance or accident, but according to the dictates of divine providence; as St. Augustine writes elsewhere, a creature attains its corporeal manifestation "when it ought to come into being."[11] Such, in briefest outline, is the traditional doctrine, at once metaphysical and theological, regarding the origin of life on earth, its authentic "evolution."[12]

It is time now to examine the teaching of Teilhard de Chardin, beginning with the subject of evolution, which proves to be central to his entire doctrine. Everything rests upon that conception. For Teilhard de Chardin, evolution is not simply a scientific theory, but an established and henceforth irrefutable truth. It is in fact the rock upon which he would found his entire doctrine. "Is evolution a theory, a system or a hypothesis?" he writes. "It is much more: it is a general condition to which all theories, all hypotheses, all systems must bow and which they must satisfy henceforward if they are to be thinkable and true. Evolution is a light illuminating all facts, a curve that all lines must follow."[13]

What is quite amazing, however, is that nowhere does this prolific author explain with even a modicum of precision on what basis he is putting forth these sweeping claims. We are told dogmatically that evolution is an established fact; but we are never told who has established it, and by what means. We are told, often enough, that

---

10. *De Trinitate*, 3.9.16.
11. *De Genesi ad Litteram*, 1.2.6.
12. It is worth pointing out that this perennial teaching comes to us from a double source: it derives not only from the Judeo-Christian revelation, but pertains to the metaphysical traditions of mankind as well. See especially Ananda Coomaraswamy, *Time and Eternity* (Ascona: Artibus Asiae, 1947), a masterful study which documents this truly universal doctrine based on Hindu, Buddhist, Greek, and Christian texts.
13. PM, p. 219. See List of Abbreviations, p. 256.

the doctrine is founded upon evidence, and that indeed this evidence "is henceforward above all verification, as well as being immune from any subsequent contradiction by experience"[14]; but we are left entirely in the dark on the crucial question wherein, precisely, this evidence consists.

In *The Phenomenon of Man*, it is true, Teilhard leads us to believe at one point that the doctrine of organic evolution has been confirmed by paleontological findings. When, in 1859, Darwin first promulgated his discovery, Teilhard admits, the theory had not yet been adequately verified. "But things are now changing. Since the days of Darwin and Lamarck, numerous discoveries have established the existence of the transitional forms postulated by the theory of evolution."[15] But it happens that this statement is misleading, to say the least. For as George Simpson (an ardent evolutionist, and one of the foremost authorities) points out: "It remains true, as every paleontologist knows, that most new species, genera, and families, and that nearly all categories above the level of families, appear in the [paleontological] record suddenly, and are not led up to by gradual, completely continuous transitional sequences."[16] One also knows, moreover, that this circumstance has been from the start one of the major stumbling blocks for the proponents of evolution: the fossil record is definitely unfriendly to their cause. So much so that, as we have noted before, evolutionists have all along been obliged to postulate *ad hoc* hypotheses of even the most farfetched kind in order to explain away the persistent and ever-recurring difficulty of "missing links." To be sure, the Jesuit paleontologist knows this very well, and is in fact himself the originator of one of these ingenious postulates (the so-called "automatic suppression of origins"[17]). But then, if the expected transitional forms have not been discovered, and if, according to an hypothesis which he

14. Ibid., p. 140.
15. Ibid., p. 82.
16. G.G. Simpson, *The Major Features of Evolution* (NY: Columbia University Press, 1953), p. 360.
17. According to this doctrine (presented by Teilhard de Chardin before the Congress on Philosophy of Science, Paris, 1949), the birth of a new phylum is accomplished in a short period of time through a small number of individuals, all

himself has introduced, they cannot be found at all, why does he inform his readers that "numerous discoveries have established the existence of the transitional forms postulated by the theory of evolution"? In the sense, at least, in which this statement is sure to be read, it is not only misleading, but undeniably false.

It has sometimes been pointed out by philosophers of science that the theory of evolution is not in fact empirically based. Even Ernst Haeckel, the celebrated Continental evolutionist (reputed to have been "more Darwinian than Darwin himself") must have been aware of this when he wrote to a scientific friend that "one can imagine nothing more absurd, nothing which indicates more clearly a total lack of comprehension of our theory, than to demand that it be founded upon empirical evidence."[18] Now it may come as something of a surprise that Teilhard himself implies as much sixty pages later in *The Phenomenon of Man*. "In view of the impossibility of empirically perceiving any entity, animate or inanimate, otherwise than as engaged in the time-space series," he writes in a footnote, "evolutionary theory has long since ceased to be a hypothesis, to become a (dimensional) condition which all hypotheses of physics or biology must henceforth satisfy."[19] In other words, we are now led to believe that the truth of evolution can be established, not by direct empirical evidence, but somehow, Kantian style, on *a priori* grounds, by analyzing the conditions to which all observation must submit. And this, too, must be the reason why the truth of evolution is said to be "above all verification," and why it is "immune from any subsequent contradiction by experience." It is "above all verification" because it cannot be empirically verified at all.

Everything, then, appears to hinge upon Teilhard's "Kantian argument," or rather, upon the question whether that argument is correct. If it is, then the issue has indeed been settled once and for all, as Teilhard insists. It is quite surprising, therefore, that this crucial

---

of slight stature and composition, which consequently disappear without leaving any trace in the fossil record.

18. Quoted by Louis Bounoure in *Déterminisme et Finalité* (Paris: Flammarion, 1957), p. 48.

19. PM, p. 140.

discovery has been relegated to a footnote and mentioned only in passing, as it were. And it is surprising, too, that it should have been left to the reader to ascertain why the impossibility of perceiving physical entities "otherwise than as engaged in the time-space series" leads ineluctably to the conclusion that "organic evolution exists, applicable equally to life as a whole or to any given living creature in particular."[20]

Nonetheless, let us try to understand what Teilhard is suggesting, however obscurely. The point seems to be that inasmuch as the emergence of living forms can be "empirically perceived," this genesis can be conceived or imagined in none other than transformist terms. Now this could well be true. But the crucial difficulty lies precisely in the initial premise: in the postulate, namely, that the emergence of living forms *can* be "empirically perceived." First of all, one knows that such a perception has never taken place. As Jean Rostand points out: "We have never been present even in a small way at *one* authentic phenomenon of evolution";[21] which is to say that no one has ever observed a *bona fide* transformation of species, be it by direct or indirect means. Furthermore, there is absolutely no reason to suppose that everything within Nature *can* in fact be "empirically perceived," and that all happenings without exception can be neatly parcelled out in what Teilhard is pleased to call "the time-space series." This is not an *a priori* condition at all, but simply an unwarranted assumption. What it amounts to, basically, is that one limits the possibilities of the real by postulating that in principle it must fall entirely within the scope of human observation. But we know for a fact that this is not true: if it were, there could, for example, be no such thing as an electron (seeing that an electron is both a particle and a wave, or better said, is neither of the two, and is therefore definitely *not* "empirically perceivable"). The point is that neither the perceivable, nor the much larger category of the imaginable, can cover the entire ground of even physical reality.

It thus turns out that Teilhard's Kantian argument in support of

---

20. Ibid.
21. *Le Figaro Litteraire*, April 20, 1957. Quoted by Titus Burckhardt in *The Sword of Gnosis,* ed. J. Needleman (Baltimore: Penguin, 1974), p. 143.

evolution is spurious: its central premise proves to be simplistic and naïve. At bottom his reasoning reduces to a confession of incapacity: like Jean Rostand, he firmly believes in evolution "because I see no means of doing otherwise."[22]

It may be that Arthur Koestler has somewhat overstated the case when he referred to Darwinism as "a crumbling fortress"; however, one cannot but agree with Ludwig von Bertalanffy (a distinguished biologist, let us add) when he writes: "The fact that a theory so vague, so insufficiently verifiable, and so far from the criteria otherwise applied in 'hard' science has become a dogma can only be explained on sociological grounds."[23]

Yet Teilhard, to be sure, perceives the matter otherwise. "Using the word 'evolution' in its most generally accepted meaning, and in a purely experiential context," he writes, "I would say that man's origin by way of evolution is now an indubitable fact of science."[24] Indeed, the point is made in the most emphatic terms: "There can be no two ways about it: the question is settled—so finally that to continue to debate it in the schools is as much a waste of time as it would be to go on arguing whether or not the revolution of the earth is an impossibility." We are here confronted by a strange phenomenon: what should one say? How can an informed and seemingly rational person make such misleading claims? Let us consider the matter.

To begin with, it is to be noted that the concept of organic evolution, as it is generally understood, not only postulates the occurrence of macroevolutionary transformations, but also entails certain assumptions regarding the *modus operandi* by which the conjectured transformations have been accomplished. For indeed, even though evolutionists may disagree about many things, and although

---

22. S. Stanley, *Macroevolution* (San Francisco: Freeman, 1979), p. 39.
23. Quoted by Huston Smith in *Beyond the Post-Modern Mind* (NY: Crossroad, 1982), p. 173.
24. SC, p. 139.

they may no longer be Darwinists in the full sense, it is still generally believed that the origin of species comes about, not by design, nor by the operation of final causes, but by means of a random process. Life emerges and unfolds its myriad forms by accident, or as we say, by chance: that is the fundamental idea.

But what does this actually mean? What *is* chance, after all? The notion turns out to be inherently negative: to say "chance" is no more than to deny that the event in question has a sufficient reason, that it fulfills a purpose or design. With reference to the origin of species, moreover, this is tantamount to the denial of any efficacious creative act. So long as it is admitted that these origins conform to the will or the plan of God, one can no more speak of natural selection (or of any other stochastic means) than one could speak of probabilities with reference to a card game in which the deck has been stacked. At bottom, therefore, Darwinism amounts to no more and no less than a denial of God's creative efficacy in the sphere of biogenesis.

It may not be without interest to point out that Darwin's own road to discovery confirms the atheistic underpinning of his doctrine. One knows from his early notebooks that he was a thoroughgoing materialist who had ruled out from the very start the possibility of any divine creative act; and clearly, as Stanley Jaki has observed, "Darwin could not proceed along lines of inductive reasoning in the *Descent* if he had reached its main conclusions more than thirty years earlier."[25] Where God is denied, the origin of species must indeed be viewed, not only in transformist, but more specifically in statistical terms: under such auspices Darwinism of some kind becomes the only thinkable option. One has then no choice but to accept the Darwinist postulate of organic evolution, despite the "astronomical" unlikelihood of its claims. As James Gray (himself an ardent evolutionist) has said: "No amount of argument, or clever epigram, can disguise the inherent improbability of orthodox evolutionary theory; but most biologists feel it is better to think in terms of improbable events than not to think at all."[26]

---

25. *Angels, Apes, and Men* (La Salle: Sugden, 1983), p. 53.
26. Quoted by Stanley Jaki, op. cit., p. 65.

But let us get back to Teilhard de Chardin. Unlike Darwin, the French Jesuit was certainly not an outright materialist: it appears that he wanted to have God and evolution too. Or better said, would have evolution, and as much of God as that concept will permit. And let there be no mistake about it: for Teilhard evolution was far more than simple transformism in the minimal sense. There can be no doubt that when it comes to what he takes to be the earlier stages of evolution (right up to the first appearance of man), his vision of the evolutive process is Darwinist to a high degree. Not only, thus, does Teilhard speak repeatedly of such things as "happy accidents" and "means of survival," which are then "promptly transformed and used as an instrument of progress or conquest,"[27] but he goes so far as to propound a theory of "groping" as "the specific and invincible weapon of all expanding multitudes."[28] At one point he seems to differentiate his own theory from the classical Darwinist conception by telling us that it would be a mistake to interpret "groping" as mere chance. "Groping is *directed chance*. It means pervading everything so as to try everything, and trying everything so as to find everything."[29] But the difference turns out to be slight. For as Teilhard himself explains later on: "It is only really through strokes of chance that life proceeds, but strokes of chance which are recognized and grasped—that is to say, psychically selected."[30] What this amounts to is that, in addition to "natural selection," Teilhard postulates "psychic selection" as yet another evolutionary mechanism. Meanwhile it remains true, in the new doctrine no less than in the old, that life proceeds "only through strokes of chance."

Whether or not Teilhard regarded this additional article of evolutionist belief as having been validated by what we have called his Kantian argument, it is clear in any case that he presents what might

---

27. PM, p. 104.
28. Ibid., p. 110.
29. Ibid.
30. Ibid., p. 149.

be termed the stochastic postulate as nothing less than an irrefutable truth of science. That evolution proceeds "only through strokes of chance" has become for him a fundamental dogma. And he insists that the discovery of this chance-driven evolution constitutes in fact the decisive event of modern times: "What makes the world in which we live specifically modern is our discovery in it and around it of evolution."[31] We are told, moreover, that in consequence of this fact, all basic conceptions must now be reformulated in essentially evolutionist terms: in this modern world, dominated by the discovery of evolution, all time-honored beliefs are to be rethought and rectified, Teilhard maintains. What he imagines to be "the truth of evolution" is for him the touchstone by which all human conceptions are henceforth to be tested. And not only human truth, but all that earlier generations had held to be a sacred and more-than-human truth. "What we have to do without delay," Teilhard tells us in particular, and in no uncertain terms, "is to modify the position occupied by the central core of Christianity...."[32]

The first conception that needs to be reformulated is the Judeo-Christian idea of creation. We are told that it is no longer possible to accept what Biblical revelation, as interpreted by two thousand years of Christian tradition, has to say concerning the creation of the world and the origins of life. It appears that Bible and Tradition must henceforth bow before Darwin's *Origin of Species* and *The Descent of Man*. Not that we need to deny God or reject the idea that He has created the world: Teilhard's doctrine is a good deal more subtle than that. Certainly he does not align himself in this regard with Darwin and the atheistic materialists. What needs to be done, he believes, is recast the idea of God and the conception of His creative act so as to conform that Christian teaching to the magisterial truth of evolution. Teilhard thinks that hitherto Christianity has been too generous in its estimate of what God can do: it has grossly overrated His creative efficacy. God does create, he assures us—but not in the absolute sense of traditional theology. God creates, but *only by way of evolution*: that is the new dogma. We are told explic-

---

31. Ibid., p. 229.
32. CE, p. 77.

itly that "God cannot create except evolutively"[33]: that is what we must henceforward believe.

But why? To begin with, there is in truth absolutely no reason to assume that the "evolution" to which God is supposedly restricted exists in the first place: enough has already been said to establish this point. But even if evolution could indeed be substantiated as a scientific theory, this still would not provide a sufficient basis upon which to challenge the traditional Christian doctrine of creation, let alone found a new theology. And the reason for this insufficiency, clearly, lies in the fact that *qua scientific theory* the doctrine is strictly confined to the realm of phenomena: it then speaks only of things that can in some sense be empirically observed, and only insofar as they *can* be thus observed. But this obviously excludes from consideration not only God, but His creative Act. And even Teilhard admits that "where God is operating it is always possible for us (by remaining at a certain level) to see only *the work of Nature*," and that "we shall never escape *scientifically* from the circle of natural explanations."[34]

But then, if that be the case, by what right does he maintain that the traditional Christian doctrine, which is in the first place theological and metaphysical, has in fact been invalidated by the discovery of evolution? If science is unable to penetrate beyond the level of phenomena to behold the secret working of God, how can it enlighten us in that regard? At best it can affirm that the phenomena do not suffice, that the pieces do not constitute a coherent whole, and that consequently (on the strength of some categorical imperative) there *must* be something above and beyond the phenomena: a factor X, which by virtue of its transcendence remains forever unknown and unknowable. The point is that science may intimate that God exists, but cannot enlighten us further on theological issues. It can suggest that God created the world (as a habitation for man, no less), but can tell us absolutely nothing regarding the *modus operandi* (if one may put it so) of the creative act. Nothing, therefore, obliges us to conclude that "God cannot create except

33. Ibid., p. 179.
34. SC, pp. 27–28.

evolutively." Whatever may have been the train of thought by which Teilhard arrived at this remarkable notion, it is clear that the dogma has no basis whatsoever in scientific fact.

This brings us to a key question: what is it in the idea of evolution which so greatly fascinated and inspired our Jesuit paleontologist? Is it simply the notion that lizards have descended from fish, mammals from lizards, and man from primate stock? As Teilhard himself informs us in a most interesting piece entitled "Note on the Essence of Transformism," that is not really the point at all. Transformism, in its true and essential sense, he tells us, is far less specific than that. What it really means is that the origin of life and of species can be adequately understood in terms of physical causes, or in terms of a "physical connection." Surprisingly enough, however, that "connection" need not necessarily be understood in terms of filiation or lines of descent. "Without as yet pre-judging in any way the particular physical nature of this connection," he explains, "and without even asserting that there is a line of descent, properly so called, linking organic beings, we hold firmly to the belief that the various terms of life appear as a physical response to one another."[35] At the same time Teilhard is careful to point out that the idea of transformism, thus understood, does not exclude the notion of a divine creative act: "For the transformist retains the right, as much as anyone else, to believe that a creative act is necessary to set the world in motion. What he postulates is quite simply that this perennial and indispensable act on the part of the first cause comes to us in the order of history and experience in the form of an organically established movement."[36] Now this, of course, can hardly be denied. What we encounter "in the order of history and experience" is indeed an organically established movement: what else could it be? Is this, then, what the transformist postulates? Obviously, there must be more to the transformist hypothesis; and this more, quite

---

35. HM, p. 110.
36. Ibid., p. 112.

clearly, must lie in the idea that the network of physical causes is in itself sufficient to explain the phenomena of biogenesis and species formation. Or to put it another way: the point of transformism, as Teilhard presently informs us, is finally to rule out "the intervention of an extra-cosmic intelligence"[37] in the operations of Nature.

One might remark that there is something distinctly Newtonian or Cartesian in this notion of all-sufficient causality. One has the impression that Teilhard stands yet under the spell of the Newtonian world-picture, with its now antiquated materialist conceptions. His notion of an unbroken network of physical causes in terms of which everything can be deterministically accounted for seems to belong to an era in which the principle of "complementarity" in the sense of Niels Bohr had not yet come to light, and the sobering recognition that our scientific knowledge of Nature is perforce fragmentary had not yet dawned upon mankind. It is in the spirit of this unrefined and unchastened *Weltanschauung* of a bygone era that Teilhard envisages Nature as a seamless garment of organic interrelationships, in which every living form can be regarded as the resultant of preceding forms. Thus conceived, transformism is in essence the biological counterpart of Newtonian physics, and the biosphere an analog of that "clockwork universe" which had been the dream of the classical physicist.

There must be a rigorous physical connection, a "physical agent" as Teilhard says, in terms of which biogenesis and speciation can be explained; but what exactly this connection, or this agent, might be is another question. What alone is essential to the transformist cause, he believes, is that there be such a connection, such a physical determinism. This is what every "natural scientist worthy of the name" assumes: "He may hesitate about the precise nature of the physical agent shared by the successive forms of life; but the belief that such an agent exists, whether it be confused with the generative function or not, the dream that one day we shall be able to put a name to it and define its behavior, it is there we find his most precious conviction and his grandest hope."[38]

37. Ibid., p. 113.
38. Ibid., p. 112.

It is worth pointing out that this passage flatly contradicts what Teilhard has so often and so vehemently affirmed elsewhere: for if it has not yet been established that the postulated "agent" resides in the generative function, how can it be said that evolution, in its "most generally accepted sense," must henceforward be regarded as an "indubitable fact of science"? If almost a century after the publication of *The Origin of Species* scientists are still dreaming that someday they will know whether or not there exist actual lines of descent connecting higher to lower forms of life, how then can one say that the matter has been settled "so finally" that it would be a "waste of time" to discuss it further?

One wonders, too, what it is about the transformist hypothesis that inspires such strong convictions and powerful sentiments in its votaries. Why does the transformist believe or dream with such vehemence? From whence does he derive "his most precious conviction and his grandest hope"? It appears to be an act of faith, as Teilhard himself suggests: "It is 'faith' in one organic physical interaction of living beings, it is that and *nothing else,* which constitutes the necessary and sufficient disposition for an evolutionist mind."[39] But the question remains: what is the basis of that faith? And why does the issue loom quite so large? Why this almost religious intensity? The reason, as we would like now to point out, is that the issue *is in fact religious to the core.* Teilhard himself, moreover, makes this abundantly clear: "We have to make a choice: there is either evolution or intrusion," he declares. And what might be the nature of this "intrusion"? It is precisely "the intervention of an extra-cosmic intelligence." This, then, must be the crux of the matter: the essence of transformism reduces ultimately to a denial of God's role or efficacy in the generation of living forms. In the final analysis, this is what Teilhard perceives as the quintessential faith of the evolutionist, "his most cherished conviction and his grandest hope"!

At first glance this may strike us as rather incongruous: was it not the express intention of the Jesuit paleontologist to reintroduce the Christian God into the scientific world-view? Was it not the great ambition of his life to demonstrate that the idea of evolution, in its

---

39. Ibid., p. 111.

full-blown form, demands actually a Christic "Point Omega" as a universal center of attraction and confluence? Now this assessment is almost correct; it is right except for one crucial point: the God to be installed upon the evolutionist horizon is no longer the traditional Christian God, the supposedly "extrinsicist" and "immobilist" God of a pre-scientific humanity. It had never been Teilhard's intention to defend and reinstate the traditional Christian teaching; instead, his objective from the start has been to *reshape* the doctrine. "What we have to do without delay is to modify the position occupied by the core of Christianity...." In a word, Teilhard's objective is to found a new Christianity. And that must be the reason why he is willing to join forces with the materialists, the atheistic Darwinists, in their well-camouflaged campaign against "interventions by an extra-cosmic intelligence." The thrust is aimed, not at the new, but at the old religion: at traditional Christianity, with its time-honored belief in an eternal, transcendent, and absolutely omnipotent God. It is this "God of the Above" who needs to be overthrown to make way for the new deity. "This is still, of course, Christianity and always will be," Teilhard assures the reader; "but a Christianity re-incarnated for a second time in the spiritual energies of Matter. It is precisely the 'ultra-Christianity' we need here and now to meet the ever more urgent demands of the ultra-human."[40]

To be sure, whether "this is still, of course, Christianity" is another question, an issue to be dealt with in the final chapter. What concerns us at the moment is the recognition that a new or "reincarnated" religion has appeared upon the scene, and that Teilhard is its self-appointed prophet. And this I deem to be the decisive fact, the key to the phenomenon of Teilhard de Chardin.

---

40. Ibid., p. 96.

2

FORGOTTEN TRUTHS

ACCORDING TO Teilhard's usage of the word, "spirit" is more or less synonymous with "consciousness" and "thought." And it is scarcely surprising that he thinks of spirit in evolutionist terms: "From a purely scientific and empirical standpoint, the true name for 'spirit' is 'spiritualization.'"[1] It is a process, then, "a gradual and systematic passage from the unconscious to the conscious, and from the conscious to the self-conscious."[2] Now this second mode of consciousness appertains to the human sphere: it is a characteristic of man. We humans not only know, but know that we know; and let us add that this reflective mode of awareness is evidently connected with the specifically human phenomenon of language. Once symbols have taken the place of things, one is able to manipulate these symbols, move them about at will; and by virtue of this inner freedom one finds oneself, as it were, in a new space: the space of concepts, the inner world of thought.

Consciousness and thought: this is what Teilhard is referring to when he speaks of "spirit." There is a birth or genesis of spirit, moreover, because there is an emergence of consciousness and thought. Spirit has become in effect a phenomenon—an "internal" phenomenon, perhaps, one that, strictly speaking, can be "observed" only in ourselves; but still a phenomenon. And as such it is related to other phenomena: to biological complexities, and to brain function. At Teilhard's hands spirit becomes a variable correlated to other variables, other parameters.

1. HE, p. 96.
2. Ibid.

Spirit, then, is growing; it is evolving via a more or less continuous trajectory, along which it is possible to distinguish various stages, ranging from the most dim and rudimentary consciousness in the lowest forms of life right up to the peaks of human thought. And beyond this, what for us unimaginable heights of thought, or hyper-thought, may yet be attainable in the far distant future if our individual powers become more and more amplified through ever-greater strides of socialization? What if Teilhard is right in conjecturing the formation of a super-organism, a Gargantuan creature endowed with a brain within which each of our brains is but a single neuron as it were?

But let us take spirit at its present high-water mark: it is, then, thought. This is Teilhard's position; and it is not without interest to note that this thesis has a distinctly Cartesian ring. It was Descartes, as we know, who implicitly identified the inner man, the spiritual man, with thought: *"cogito ergo sum."* It was Descartes who thus reduced spirit to a *res cogitans,* a "thinking entity." Until then, spirit (in the form of intellect) had been distinguished from "mind" or the discursive level of psychic function. And this fundamental distinction had been clearly marked in the ancient metaphysical vocabularies: it is the difference between *nous* and *dianoia* in Greek, *intellectus* and *mens* in Latin, or *buddhi* and *manas* in Sanskrit, to mention the most important examples. But with the advent of the modern age, the two levels of cognition came to be *de facto* identified; or what amounts to the same, the higher of the two was no longer discerned. Men forgot all about "intellect" in the lofty and time-honored sense of that term, to the point where a so-called "intellectual" is someone who denies that there could be such a faculty in man. It is one of the signs of our time that the "intelligentsia" has become hostile to intellect.

But these are matters to be dealt with later. For the moment it suffices to say that Teilhard, for his part, conforms to this modern trend, and seems to accept its implicit presuppositions without the slightest qualms. In fact, he goes considerably further than the Cartesians: for whereas Descartes still believed in "mind" as a spiritual substance, a *res cogitans* whose activity is thought, Teilhard veers towards the position that mind or spirit *is* thought. His outlook represents thus a decisive step beyond the *cogito ergo sum*: not only is thought an indication that I exist as a spiritual entity, but it *is* that entity. According to this view the "I" is but the epicenter of the thought; what else could it be from a radically evolutionist point of view?

The traditional picture, of course, is quite different. In light of traditional philosophy one can say that thought is an activity arising from the interaction of spirit (or soul) and body. It could thus be compared to the music produced when a pianist plays upon his instrument. There is of course a correlation between the sound and the movement of the keys and hammers; and in a sense it is the piano that produces the sound. And yet it could not be said that the music is an "epiphenomenon" of the instrument, or that the two constitute complementary aspects of one and the same underlying reality or process (a view which would correspond roughly to the position of Teilhard de Chardin). The point is that in either case we have left out of account a vital factor, in the absence of which there can be no music at all: the pianist, namely. So too it happens that the brain, though necessary for thought, is not in itself sufficient to produce that phenomenon. As Wilder Penfield, a noted neurologist and brain surgeon, has put it:

> Because it seems to be certain that it will always be quite impossible to explain the mind on the basis of neuronal action within the brain, and because it seems to me that the mind develops and matures independently throughout an individual's life as though it were a continuing element, and because a computer (which the brain is) must be operated by an agency capable of independent understanding, I am forced to choose the proposition that our being is to be explained on the basis of two fundamental elements.[3]

It is clear, moreover, that the second of these two fundamental factors could not be *thought*, which is, after all, an effect: the result of the postulated interaction. It must therefore be something else: an unknown, a factor X, which we may call "mind" "soul," or "spirit," as

---

3. *The Mystery of the Mind* (Princeton University Press, 1975), quoted in E.F. Schumacher, *A Guide to the Perplexed* (NY: Harper, 1977), p. 76.

we wish. In the words of Sir Charles Sherrington, it "goes in our spatial world more ghostly than a ghost. Invisible, intangible, it is a thing not even of outline, it is not a thing."[4] No wonder the renowned neurophysiologist was forced to concede that science "stands powerless to deal with or to describe" that elusive element.

Yet it exists and is no doubt the crucial factor: the active principle, the determinant of our thought. And by the same token, is this not, too, the decisive factor which spells the difference between one man and another? Are we to suppose that the difference between a Mozart and an average man is simply a matter of neurons? Are not their brains very much alike? And if there be some differences as regards, say, anatomical structure, could *this* be the cause on account of which one is a musical genius and the other is not? To say so, clearly, would be to fall once again into the old materialist position which Wilder Penfield has rejected: it would be to assume that the workings of the mind can be explained on the basis of neuronal action within the brain.[5] Let us admit it once and for all: the mind cannot be thus explained. It cannot be "explained away," in other words; there exists an inner man, a spiritual man, as religion has always proclaimed, and science "stands powerless" in the face of that spiritual being.

What, then, can we say concerning that inner man? We maintain that he is not the body, but an "indwelling spirit"; he is not the thought, but the author of the thought, the thinker. We do not know this man directly: he is invisible to our senses; but yet we *do* know him. And this is perhaps the greatest miracle of all: he is no stranger to us. We know him by his words, we know him by his actions, we know him by his countenance, by the expression on his face and the look in his eyes. After all, the body is *his* body, his instrument—his *icon*, one is tempted to say.

Now this is the simple and eminently natural doctrine which mankind had always accepted—until just yesterday, when it was replaced within the ranks of the modern intelligentsia by a scientistic

---

4. *Man On His Nature* (Cambridge University Press, 1951), p. 256.
5. I have dealt with this issue at length in "Neurons and Mind." See *Science and Myth* (Tacoma, WA: Sophia Perennis/Angelico Press, 2012), chap. 5.

teaching which explains nothing, and for which there is finally not a shred of authentic evidence.

What Teilhard has failed to grasp is that spirit and matter are situated on different levels of reality. They are not simply two faces of a single cosmic substance or principle, but two tiers of the cosmic edifice: the two poles, one could say, between which the entire drama of cosmic existence is played out.

There is in reality no single cosmic substance, no "stuff of the universe" in Teilhard's sense. That is a pure fiction, a carryover from the old Newtonian materialism which science itself has since been forced to disavow. Now it is true, certainly, that we cannot stop at the idea of multiplicity, or of duality: our intellect craves unity, it would know "the One." But as all the traditional schools of metaphysics have recognized, that One is not a part of the cosmos, nor is it the cosmic whole (what our contemporaries would call "the holistic universe"): instead, it is the Absolute and the Infinite, which are none other than God.

The great fact is that the world begins with a primordial duality: "*In principio creavit Deus caelum et terram.*" Theologians do not always agree in their interpretation of this "heaven and earth"; and no doubt the expression can be legitimately interpreted in more than one way. What is beyond dispute, however, is the fact of duality: after the One comes—not three or four—but precisely *two*.

We have said that the Biblical "heaven and earth" can be understood in a number of senses: as the active and passive principles of cosmogenesis, as *form* and *matter* in the Scholastic sense, as the angelic and the corporeal orders of existence, as soul and body, as the male and female principles, and so forth. But it will also be noted that all these connotations are somehow related: they constitute so many exemplifications, actually, of a single underlying idea. Spirit and matter, let us say; the terms are well suited, on account of their deep and manifold associations, to convey some slight intuition, at least, of what is in truth at stake.

There are, then, these two poles, between which the entire gamut of cosmic existence is spread out—but not of course spatially, but in an ontological sense. And yet we can hardly avoid the temptation to spatialize even the purest of relations: we are never at peace until we have discovered an icon, a visual representation of the ideal or the metaphysical. Now this, of course, is exactly what the ancient cosmologies—such as the much-maligned Ptolemaic worldpicture—were attempting to do: their true function, quite clearly, was to provide a symbolic representation of the cosmos in its ontological entirety. And as we have pointed out elsewhere,[6] it could hardly have been an accident that Europe began to lose its metaphysical sense at precisely the time of the so-called Copernican revolution. Most assuredly, there is more—incomparably more!—to the Ptolemaic *Weltanschauung* than meets the modern eye. Let us not forget that the very terms "heaven" and "earth" in which the Biblical revelation apprises us of the primordial duality are spoken from a distinctly Ptolemaic point of view. Their direct reference is to a natural icon, an icon which we behold, not with telescopes or Geiger counters, but with our God-given eyes. And how wonderful, how infinitely expressive that icon proves to be!

But there exists yet another image, another icon, which is equally Biblical; for it has also been said that Heaven lies "within." Here the standpoint has changed: from an authentically metaphysical perspective, the two "worlds" are no longer separated, but interpenetrate, one could say. Yet here also we need to go beyond the image, beyond the figure of speech. Spirit and matter interpenetrate, no doubt; but they do so without mixture or confusion: they interpenetrate in a most marvelous way!

What we actually perceive, moreover, and what alone we know in our earthbound existence is neither matter nor spirit as such, but the effect or offspring, rather, of their enigmatic union. The corporeal world does not in fact exist in isolation from the spiritual, nor is it intelligible in its own right: to think of it thus is to fall prey to an illusion. And of course, one does often enough think of it in these

---

6. *Cosmos and Transcendence* (Tacoma, WA: Sophia Perennis/Angelico Press, 2008), pp. 144–149.

terms; the fallacy is implicit in the very conception of the physical universe (which goes back to Descartes). One knows today, after centuries of philosophical debate, that the Cartesian notion of a physical universe (made up of self-existent *res extensae* or "particles") is thoroughly untenable; and is this not indeed the reason why the physicist's "matter" has proved chimerical? For as one knows today on strictly scientific grounds, there actually exists no such thing as a "particle" in the classical sense. To those who believe in the so-called physical universe, this means that the cosmos itself is void not only of qualities, but of substances as well. Yet in reality the cosmos is far from void. What has happened is that we have lost our grip on reality: we have systematically "filtered out" the real.[7]

The universe is not in fact an empty space-time, whose curvatures and singularities conjure up an appearance of solidity. The material (or corporeal) world does have a content, which however turns out to be immaterial and non-corporeal in its own right: for it derives from that enigmatic interpenetration—that "conjugal embrace of the spirit"—from which the cosmos receives its fecundity. What is more: it is precisely this "spiritual content" that renders the universe intelligible. "The most incomprehensible thing about the universe," Einstein once said, "is that it is comprehensible." Yes, on the basis of physics alone, this *is* incomprehensible. Let us understand it well: *to be is to be intelligible*. What we know in all things is spirit—or spirit reflected in matter, to be more precise. That is what matter does: it reflects the spiritual light. And that light is again perceived, not by matter, but by a spiritual faculty: the mirror reflects, and the eye perceives.

According to our astrophysical cosmology, the cosmos consists mainly of empty space. Yet in truth there exists no such thing. What appears as a void when viewed through telescopes and Geiger counters turns out to be a plenum, a perfect fullness, when perceived with a spiritual eye. For in the final count, what is not perceptible to the spiritual eye does not exist. And one should add: what *is* perceived is itself spiritual. As St. Maximus explains: "The

---

7. I have dealt with this issue repeatedly. See especially *The Quantum Enigma* (Tacoma, WA: Sophia Perennis/Angelico Press, 2005).

whole of the spiritual world appears mystically represented in symbolic forms in every part of the sensible world for those who are able to see."[8] But we must not think that this perception, to which the Saint alludes, is reserved only for mystics of high rank. In its purest form, most certainly; but let us note that all of us, whether we realize it or not, have yet some share in that vision: for it is this, and this alone, that makes us human.

Such are the rudiments of traditional cosmology. The secret of Nature, the alchemy of its innermost operation, hinges upon an ontological duality: a pairing of spirit and matter. And this pairing defines a metaphysical dimension above the corporeal. There exists a hierarchic order: there are "worlds within worlds." And it happens that in this, for us unimaginably vast hierarchy of ontological strata, "our world" occupies in fact the lowest rank.

For Teilhard, on the other hand, that world—that single ontological domain—constitutes the cosmos in its entirety. More than that: it includes even God Himself! "All that exists is matter becoming spirit"[9]: these words express the quintessence of his thought. And this means that in Teilhard's eyes, matter and spirit are situated on one and the same plane: they constitute but two faces of a single cosmic reality. As Teilhard himself observes, "There is neither spirit nor matter in the world; the 'stuff of the universe' is spirit-matter."[10]

The cosmos has thus been flattened, reduced to a single plane. We must understand that primordial duality and metaphysical verticality go hand in hand. Whosoever, therefore, denies this primordial duality has in the same breath denied the concept of an *axis mundi*. There is then no more Jacob's Ladder, and presumably no more "angels of God" to ascend and descend thereon. We find ourselves then confined intellectually to this familiar universe, this

---

8. *Mystagogy*, chap. 2 (PG 91:669C); quoted in Archimandrite Vasileios, *Hymn of Entry* (Crestwood, NY: St. Vladimir Seminary Press, 1983), p. 67.
9. HE, p. 57.
10. Ibid.

"narrow world," which remains such despite what Teilhard refers to boastfully as "the discovery of Time and Space."

Admittedly our ancestors were not as well informed as we when it comes to the quantitative or "physical" dimensions of our world. By no means, however, can they be accused of "thinking small": on the contrary, it is we who are guilty of that charge. For as Huston Smith points out: "The modern West is the first society to view the physical world as a closed system."[11] Far better to think that the cosmos was created six thousand years ago while realizing that there exist higher realms, than to believe that it consists of nuclear particles resulting from a "big bang."

Getting back to Teilhard de Chardin: having abolished—or better said, *denied*—metaphysical verticality, he proceeds to find an analog, an Ersatz, within the remaining plane. This substitution constitutes in fact the salient feature of his theory: briefly stated, he has in effect replaced the *axis mundi* by the "arrow of time," which he identifies with the thrust of an imagined evolutive trajectory. To form a mental picture of this transposition, think of the integral cosmos as a three-dimensional space, in which our space-time is represented by a horizontal plane, endowed with an axis representing time. The decisive step reduces then to a projection of that three-dimensional space onto the plane which transforms the "above" into the "ahead. This is the undeclared transformation, I say, that defines Teilhard's theory in its essence: every facet of his doctrine follows from that single step. It is this hidden "rotation of axes" that permits Teilhard to falsify just about every traditional conception of theology.

Let us consider, first of all, what becomes of the conception of man, which to be sure is as central in the Teilhardian as in authentic theology: what then are the basic tenets of traditional anthropology, and how, precisely, does the Teilhardian version deviate from that foundational teaching? Now Christian anthropology, as it is com-

---

11. *Forgotten Truth* (NY: Harper & Row, 1977), p. 96.

monly understood, conceives of man as a dichotomous being, composed of body and soul. A human being, then, is made up of *two* fundamental elements, which by their union constitute the living man. And let us note immediately that this dichotomy is precisely the microcosmic exemplification of the aforementioned primordial duality: not only the cosmos at large, but man as well exist by virtue of a mysterious union, which may be characterized as an interpenetration of spirit and matter.[12]

Soul, then, is the spiritual component of man. But does this mean that soul *is* spirit, pure and simple? Not exactly. There is in fact a Christian tradition, going back to St. Paul, which clearly distinguishes the two. According to this teaching, *psyche*—that is to say, *soul* properly so called—is not the same thing as *pneuma*: is not *spirit* in the strict sense. If it be true, then, that man, as conceived in the fullness of his being, contains within himself an authentically spiritual element or component, then there must be, not two, but *three* basic elements in the human compound: *corpus, anima,* and *spiritus,* to use the Latin terms. One arrives thus at a *trichotomous* conception of man, which at first glance seems to contradict the former view.

Yet in reality—and notwithstanding the fact that the issue has at times been hotly debated—the two views are not opposed. They correspond, if you will, to alternative ways of looking at the soul. The question is whether *pneuma* belongs to the soul, presumably as its essential core, or is to be situated outside or "above" the soul. In either case one must realize that *pneuma* is, strictly speaking, a supra-formal and hence supra-individual principle: spirit is perforce something supra-human. It is, if you will, not our soul, but rather "the soul of our soul" if one may put it so. Now, what makes all this difficult and perhaps in a way incomprehensible is the fact that "above" the soul or psychic level in the strict sense, our concepts become somehow inadequate—which in fact is not surprising in the least: after all, if indeed there be such an "above," it could hardly be otherwise.

---

12. Let me note in passing that this is precisely the kind of union with which authentic alchemy is concerned.

48   THEISTIC EVOLUTION: THE TEILHARDIAN HERESY

But these, in any case, are difficult questions which we need not probe too deeply at this point. Suffice it to say that the human soul, understood as the vital and psychic principle within the human compound, depends upon another factor: the *pneuma* or spirit, namely, which both animates and enlightens it. And whether this *pneuma* be conceived as the spiritual essence of the soul or is assigned to a higher ontological plane, in either case, soul as we know it—soul as the vital and psychic principle in man—will occupy, ontologically, an intermediary position. Situated, thus, between the spiritual and the corporeal orders of existence, its function is precisely to mediate between the two realms. In a word, the soul or psyche constitutes a *mediating* principle.

This, then, is the twofold schema in terms of which the traditional anthropology has been framed. But familiar as all this may be, one tends to forget that this ontological conception has epistemological implications. Thus, when it comes to the question of knowledge—or better said, of knowing—one readily forgets that the psyche does play no more than a mediating role. We incline in fact to the erroneous view that knowing constitutes a "psychic phenomenon": that it transpires simply on a psychic plane. The decisive fact, however, is that the miracle of knowing is ultimately consummated, not on the psychic, but on the authentically spiritual plane. Spirit alone, in the final count, constitutes the primary intellect: the true "eye" by which we see, whereas the psyche, properly so called, is but a means or medium: the "glass," as St. Paul affirms, through which now we see "darkly." And let us clearly understand that spirit as such is to us invisible, not because it is remote, or perhaps does not exist at all, but for the very opposite reason: it is far too near and far too real to become an object of the mind.

But needless to say, Teilhard does not distinguish between spirit and psyche, or what amounts to the same, between intellect and mind. Having deftly disposed of the "vertical axis," there is no more spirit, and hence in truth no more psyche or mind as well: the mediating principle, after all, presupposes the extremes. Strictly speaking,

Teilhard has no concept of psyche, but only of an imagined psychogenesis. Psyche has thus become a product of evolution, which as such is itself an evolving thing, a process in its own right.

But let us get back to the traditional teaching. It happens that the categorical distinction between intellect and mind or psyche entails another, which has likewise fallen by the wayside in modern times: if, namely, the act of the intellect is cognition and that of the mind is thought, one sees that the two are by no means the same. They cannot be: they belong after all to disparate planes. One might put it this way: thought is the quest of which cognition is the end (in both senses of the term). Thought, then, is a movement; it circles around or "zeroes in" upon its object, whereas cognition is a stasis, a state of vision and of rest. Thought as such, therefore, is not a knowing, not a cognition, but at best a means. Yet it need not be, and all too often is not: thought, as we know, can fail, it can miscarry. But the same could not be said of the intellectual act, which is vision *per se*: authentic vision, no less, in which "in a certain manner subject and object become one" as Aristotle has astutely observed.

Now the connection between thought and intellect is extremely variable. Towards one end of the spectrum, thought becomes more or less automatic; it then functions with a minimum of intellectual support. In that state the brain—the computer—runs almost on its own. And this is perhaps what happens most of the time—even in the scientific domain. Yet at crucial moments, when a so-called "breakthrough" occurs, the balance shifts abruptly. Thought ceases then to be semi-automatic: it slows down, as it were, almost to the point of stasis, and for an instant turns luminous. At such moments one almost senses that another factor or principle has come into play.

Thought is a movement, we have said, and cognition "a stasis, a state of vision and of rest." But let us try to understand this more clearly. The fact is—unbelievable as it may seem—that actual cognition does *not* take place "in time." If it did, it too would be a process or a movement of some kind, which however, as we have said, it is not. Nor *could* it be: for movement involves dispersion or multiplicity, whereas the hallmark of cognition is unity. We cannot actually perceive a landscape, say, "one bit at a time," nor hear music "note by note."¹³ What is scattered in space and time is brought together in the cognitive act, which consequently cannot be successive, cannot in fact be a process, something that happens "in time."

No wonder Teilhard confuses cognition with thought: where movement or process becomes all, the die has been cast. By his radical evolutionism Teilhard has made himself unable to understand what cognition is, and wherein it differs from thought. So too can he no longer have the slightest comprehension what spirit is in its own right, and how it differs from psyche or mind. For it is clear in light of the foregoing considerations that spirit as such must transcend the temporal condition; if it did not—if, in other words, it too were a process—the same would perforce hold true of the intellective act. But as we have pointed out, it happens that the intellective act is *not* in truth successive, *not* indeed temporal: it cannot be! One is consequently forced to conclude that spirit as such is not subject to time, that it does not disperse, as it were, into a one-dimensional continuum. In the expressive language of St. Augustine, spirit is not to be counted among "the things from which times flow." It is not flux, nor does it engender flux; on the contrary, it constitutes the source whence all unity, stability and permanence to be found in the material world are derived. As the ancient metaphysicians, be they Greek or Indian, well understood, such unity as is to be found in material things is but a reflection, as it were, of a higher unity which resides on the spiritual plane.

But there is another metaphysical fact that needs likewise to be pointed out: we must not forget that spirit, too, belongs yet to the order of creation. Our reference was *not* to the Spirit of God, the Third Person of the Holy Trinity, bur only to "spirit" with a small "s," that is to say, spirit as a created reality which, symbolically speaking, may be said to constitute the "upper third" of the cosmos taken in its entirety.

We have maintained that spirit, in its own right, is not subject to

---

13. I have dealt with this issue in great detail in my chapter on "Visual Perception" in *Science and Myth* (Tacoma, WA: Sophia Perennis/Angelico Press, 2010). It turns out, surprisingly enough, that there exists hard scientific evidence in support of the stated tenet.

time; but yet, as part of the creation, it cannot be absolutely eternal either. It must consequently occupy a middle ground, so to speak, between time and eternity. Now this is what Scholastic theologians have termed "aeviternity"—a difficult notion, admittedly, but one that a Christian metaphysics cannot avoid: for the very fact of intellection, as we have seen, implies the existence of such a supra-temporal and yet "sub-eternal" state.

Yet we must also bear in mind that spirit does not simply exist by itself, in "splendid isolation" as it were, but exists rather in a certain conjunction with the material world; as the opening verse of *Genesis* affirms: "In the beginning God created *the heaven and the earth*." The two belong together: they constitute a totality, a single "organism" if you will. Aeviternity and time, therefore, are likewise coordinated; they too belong together, like the center and circumference of a circle; and as the center is one while the circumference contains a virtually infinite multiplicity of points, so it is with aeviternity and time. Despite their conjunction—notwithstanding the closeness of their union—the two are exceedingly different, to the point of exhibiting opposite characteristics. As St. Thomas Aquinas notes: "Time has before and after; aeviternity in itself has neither before nor after." But he immediately adds a highly significant clause: "Aeviternity in itself has neither before nor after, *which can, however, be annexed to it.*"[14] Here then is the crucial point: aeviternity is "oriented" towards time—as center to circumference, or heaven to earth.[15]

The question presents itself: as we ascend from the corporeal through the various levels of the psychic world towards the spiritual—as saints and mystics do in fact ascend—how far up does time extend? And if there be such a thing as a "psychic time," by what "clocks" is it measured? It will suffice us to observe, in light of traditional doctrine, that as one ascends from the corporeal to the spiritual plane, time becomes shortened or compressed relative to

---

14. *Summa Theologiae*, I, 10, Quest. 4, Art. 2.
15. On the subject of time and eternity I refer to Ananda Coomaraswamy's masterful *Time and Eternity* (Ascona: Artibus Asiae, 1947).

Earth-time, till "a thousand years" are "as one day."[16] At last a point is reached where "before" and "after" merge in a single moment, the so-called *nunc stans* or "now that stands." We know that such a "now" *must* exist: the fact of intellection demands as much. For as we have already had occasion to point out, nothing less than this could ultimately account for the miracle of knowing. If there were not within the soul an Apex beyond the flux of time, we could not perceive motion, could not become aware of change. Strange to say, only by transcending time do we recognize its existence. Contrary to Teilhardian belief, man does not fit into a temporal continuum; he is not thus spread out, not thus dismembered. Temporal by way of his body, he is supra-temporal by virtue of his intellectual soul. In the integrality of his being, man is not a time-bound creature, not a thing that "evolves": for it happens that the cognitive act which makes us human establishes the contrary.

---

16. 2 Peter 3:8.

3

COMPLEXITY/ CONSCIOUSNESS: LAW OR MYTH?

EVOLUTION, according to Teilhard de Chardin, is a directed process: it proceeds from the material to the spiritual. The basic idea is simple: "All that exists is matter becoming spirit." But the question is: how? How do material particles give birth successively to life, consciousness and intelligence? *Through complexification*, we are told. First there are scattered particles; then come atoms; then molecules; then super-molecules; then cells; then simple multi-cellular organisms; and so on up the evolutive line. And thus, by way of progressive complexification, matter gives birth to life, consciousness and thought—in a word, gives birth to "spirit." Complexification, Teilhard maintains, "is experimentally bound up with a correlative increase in interiorization, that is to say, in the psyche or consciousness,"[1] so much so that consciousness can in fact be "defined experimentally as the specific effect of organized complexity."[2]

This, in brief, is the celebrated Law of Complexity/Consciousness that stands at the heart of Teilhard's system: the entire edifice rests upon that stipulated Law, which has supposedly been put forward "from the phenomenal point of view, to which I systematically confine myself,"[3] and purports to be an empirically verifiable truth.

---

1. PM, p. 301.
2. Ibid.
3. Ibid., p. 308.

But is it? The first thing, perhaps, that should give us pause is the obvious fact that consciousness as such is not observable at all, except subjectively, that is to say, in ourselves. Each of us, presumably, perceives the world around him, and by reflection becomes aware of the fact that he perceives. We are thus conscious of the outer world, and also conscious of that consciousness, as one might say. But the point is that we are not conscious of someone else's consciousness: it is not for us "an observable." What we do normally observe are *bodies* and *behavior*. And on this basis, by way of a certain empathy, we surmise what may be going on in the mind or consciousness of another. But marvelous as this faculty may be, it is not infallible, nor does it qualify as a means of observation in a scientific sense. Even when it comes to fellow humans, therefore, it cannot be said that consciousness is an observable—what to speak of molluscs, protozoa or super-molecules!

But if consciousness is not observable, how can it be "defined experimentally as the specific effect of organized complexity"? And if only one side of the postulated equation or proportionality can be observed, how can one speak of a scientific law?

To make matters worse, it turns out that the second component in the stipulated equation or proportionality is likewise problematic: for it happens that "complexity," as Teilhard uses the term, is not in fact a well-defined parameter. It is not clear, by any means, how he proposes to define the complexity of a physical entity in a scientifically meaningful way. Could it be taken, perhaps, as the number of elementary particles, or the number of atoms, contained therein? Obviously not: for in that case a pebble would be incomparably more "complex" than an ameba. And yet—presumably for lack of a better idea—Teilhard does introduce this very notion, unpromising though it be, as a so-called "parameter of complexity for the smallest corpuscles."[4] To be sure, he is aware of the difficulty, and is careful to point out that this "parameter of complexity" cannot be applied to

---

4. MN, p. 21. The fact that this "parameter of complexity" is meaningless above the level of molecules does not deter Teilhard from drawing his so-called "curve of corpuscularization," in which the parameter in question serves as an abscissa for "corpuscles" from electrons to man!

living organisms: "Once we are past molecules," he explains, "the very hugeness of the values we meet makes any numerical calculation of complexities impossible."[5] But clearly, this is not the reason at all: it is not the quantity of atoms which makes the calculation prohibitive—as if one could not count beyond a certain number! No: one can count right up to the total number of electrons and protons in the universe. That is not where the problem lies. The difficulty resides in the fact that Teilhard's "parameter of complexity" has not the slightest biological significance, and that one has no idea how to define a "parameter of complexity" which does.

But there is yet another problem. Let us suppose that a suitable parameter of complexity can be defined; and let us assume, further, that the consciousness of living creatures, from viruses to man, can be somehow observed and registered on an appropriate scale. And let us assume, finally, that on the basis of innumerable observations it is then found that consciousness is indeed proportional to the complexity of the organisms. Would that settle the matter? Would it constitute a verification of Teilhard's thesis?

Not at all. For as we shall presently see, such a finding would accord equally well with the traditional world-view, as outlined in Chapter 2. It turns out, therefore, that the stipulated experimental confirmations do not suffice to adjudicate between the traditional and the Teilhardian positions, which means that there is more to the celebrated Law of Complexity than meets the empiricist eye. The fact is that this so-called law harbors a premise of an ontological kind, which can most readily be formulated in negative terms: its affirmation is a denial, and what it denies is the intervention of an immaterial or spiritual factor which is *not* "the specific effect of organized complexity." To be sure, Teilhard's doctrine does allow "spirit," *but only on condition that this "spirit" be somehow produced by or extracted from a material substratum.* This assumed "primacy of matter" is the crucial point, the decisive postulate. And for all his habitual equivocation, Teilhard never vacillates on this one central issue. That axiom stands above all doubt and is not debatable: from

---

5. MN, p. 47.

the start he declares himself fully determined "to avoid a fundamental dualism"[6] *One* principle, *one* matter or spirit-matter—that is unmistakably his position. If under such auspices one can speak of "creation" at all, the new teaching would read: *In the beginning God created the earth*, period.

But what exactly is wrong with a "fundamental dualism?" Why must the idea be avoided at all cost? The notion is "at once impossible and anti-scientific" we are told,[7] but we are never told *why*. It seems strange that an author as prolific and thorough as Teilhard de Chardin should be unwilling to devote even a paragraph somewhere to the explication of this crucial point. And yet the reason for this neglect is not far to seek: it happens that the rejected tenet is *not* impossible and *not* "anti-scientific" at all. In fact, if "impossible" means *inconceivable*—and what else could it mean?—then Teilhard's first contention is already refuted by the mere fact that the idea of primordial duality has been entertained by the better part of mankind for thousands of years. And as for Teilhard's second claim—the contention that the idea of a primordial duality is "anti-scientific"—even a modest background in the philosophy of science should make it clear that science as such has actually nothing to say on questions of this kind. The point is that science deals with *phenomena*, whereas the dualism in question refers to metaphysical principles. The charge that the metaphysical conception of a primordial duality is "anti-scientific" proves therefore to be illegitimate.

The stipulated Law turns out to be a Trojan horse: something quite unsuspected has been smuggled in under the cover of that so-called "phenomenal point of view, to which I systematically confine myself." As happens quite generally when in the name of Science metaphysics is officially banned, the despised discipline returns by way of the back door.

---

6. PM, p. 64.
7. Ibid.

The fact is that Teilhard has not brought forth a single cogent argument against the traditional dualism. Once the smoke of spurious dialectic and overbearing innuendo has cleared, one sees that the doctrine remains as viable in this 21st century as it was in the days of Moses, Plato, and the Sānkhya cosmologists of ancient India. The real question, in any case, is whether Teilhard's evolutionist monism stands half as well. Could it not be his own stipulated Law of Complexity/Consciousness—replete with its anti-dualist thrust—that is in fact "at once impossible and anti-scientific"?

The first thing to be observed in this connection is that consciousness cannot after all be conceived as a "specific effect of organized complexity." The claim is fraudulent. Consider, for example, the act of visual perception. The entire optical mechanism is there to translate the "information" contained in an external panorama into a particular "state" of the visual cortex, defined (let us say) by the "on" or "off" positions of a million neurons. So far the process is perfectly familiar and comprehensible to the engineer: this is what also happens, basically, in photography or television. In the case of photography, for example, the end-product of the entire operation is a piece of paper, say, covered with a fine grid of black and white (or colored) dots. What has been produced is an organized multiplicity of some sort. An actual perception, on the other hand, is something utterly different: not an organized multiplicity, but a *structured unity* one can say. The passage from a photograph—or a state of the visual cortex—to an actual perception is nothing short of a miracle: the *many* have somehow become *one*.

Let us try to understand this absolutely essential point as clearly as we can: to perceive the picture, it is necessary to take cognizance of a million dots, or a million neurons, *all at once*. But this is something that no conceivable mechanism could accomplish. All that a mechanism, be it physical or biological, can accomplish is to transform an input (be it continuous or discrete) into an organized multiplicity of some kind; for what is itself dispersed cannot produce non-dispersion.

It is really as simple as that. What is not simple, on the other hand, is to understand, even in the most cursory manner, how "non-dispersion" does come about. This is where what Wilder Penfield has termed "the second fundamental element" comes into play. And this elusive "factor X," which seemingly "reads" the cerebral computer, is something which cannot but baffle us: truly "it goes in our spatial world more ghostly than a ghost" as Sir Charles Sherrington has been forced to admit.

There is, however, an important observation to be made: the reason why the element in question—which in traditional parlance is called the soul—"goes in our spatial world more ghostly than a ghost" is that the element is not itself a spatial entity. This much, at least, we are able to understand: the soul is not a material entity, not something that admits extension or can be localized in space. René Descartes was wiser, after all, than his monist disciples when he observed that *res extensae* alone do not suffice to make a world. It turns out that not only thought, but perception as well demands a supra-spatial principle. And this is also of course the reason why organized multiplicity or so-called "complexity" does not suffice.

But Teilhard seems not to have grasped this basic point. If he had, he never could have maintained that "space-time contains and engenders consciousness,"[8] nor could he have spoken of consciousness as "the specific effect of organized complexity." What space-time contains and may conceivably engender, and what alone can be produced as the specific effect of a spatially organized complexity, is again a spatially organized complexity of some kind. But "spirit," as we have seen, happens not to be an organized complexity at all. And so we discover that it is in reality the Law of Complexity—and not the primordial duality—that proves to be impossible.

What the traditional teaching affirms is that consciousness derives, not from the body, but from the soul. Consciousness, then, is a capacity or power of the soul (a *"pouvoir"* in the sense of the biologist Maurice Vernet). And as a capacity it is independent of the body. What might be termed empirical consciousness, on the other

---

8. PM, p. 259.

hand, is obviously "somatic" in the sense that it arises from the interaction of body and soul. And this empirical consciousness, which we normally enjoy during what is called the waking state, is of course dependent, as we know, upon the body and its "organized complexities." Yet it is nothing other than the realization of a capacity or *pouvoir* which as such belongs to the soul and is given from the start. The case is entirely analogous to that of a pianist playing upon his instrument: a certain power belonging to the former is actualized. The point is that neither the latent nor the actualized power belongs to the instrument: it is definitely not the "specific effect" of the piano! Whether it be in potency or in act, the art belongs to the artist alone: it evidently does not reside in the hammers or strings.

And there is another point to be made. Souls do not—nay, *cannot*—"evolve": they are created. They come into existence, thus, not by a slow and groping process as Teilhard imagines, but instantly, *omnia simul*, in the indivisible *nunc stans* in which God created the cosmos and its "times." From an empirical standpoint, on the other hand, a certain evolution does take place. The concept has a precise traditional sense, which corresponds in fact to the etymology of the word: what exists from the start, *sub specie aeternitatis*, "unfolds" in time by way of its corporeal manifestation, beginning with its specific moment of birth. The soul, in other words, manifests itself progressively through its body. Its own body, let us add: for the connection is far from adventitious.

Our life, then, is an evolution in that sense. We are here to unfold the "talents" which have been inscribed upon our soul. And something similar can be said, no doubt, with reference to animals: they, too, have their powers, their specific "talents." Only it must be clearly understood that these differ markedly from our own. One cannot say, for instance, that the *pouvoir* of a chimpanzee includes the composing of symphonies; to think in such terms is to enter straightway into the realm of fantasy. The powers in question are quite specific: they are in fact determined by the species. Every creature acts according to its nature, and within the limits imposed by that nature. The latent consciousness in a newborn monkey, for instance, is very definitely a simian consciousness. And that is the reason why the young monkey does not take to the water like a duck; it takes to the trees instead. Naturalists call it instinct; but whatever it be called, one needs to realize that the propensity in question is of a *psychic* nature: it belongs to "the second element," the agent as opposed to the bodily instrument. Instinct, thus, is something inscribed in the latent consciousness of the creature.

There is, then, an evolution of individuals; but is there not also an evolution of species? In a sense there is. There has obviously been an evolution of mankind. Each of us has in fact been formed in part by a cultural development going back to remote times: the species, too, may thus be said to "evolve." But strictly speaking, what thus evolves is not the species as such, but its manifestation, which is to say that the stipulated evolution cannot transgress the limits imposed by the species as such. There is in that sense an evolution of species—which in fact constitutes a microevolution—but there is no such thing as Darwinist transformism, for the very simple reason that nothing can become what it is not.

Everything is comprised *ab initio* in that capacity, that *pouvoir* of life, which resides in the species. And this capacity is specific: it includes certain possibilities, and excludes others. There is really no such thing as a universal capacity: an aptitude is always a fitness for some specific function. It is a vector with a given magnitude and direction. Even the rational faculty in man is no exception: it too has its own sphere of operation, outside of which it is powerless.

The *pouvoir* of life does not derive from the body, we have said; it cannot be explained in terms of somatic complexities. Thus, even if it were possible to transmute the body of an ape, say, into that of a man, that transformation could at best produce another ape, and indeed a very sick one at that. For we must remember that the bodily form, with all its complexities, is marvelously adapted to the powers of the soul. And that is the reason why beyond a certain point, mutations of structure result—not in a Darwinian evolution—but precisely in that separation of body and soul which constitutes death.

Teilhard himself seems not always to be satisfied with the notion that life and consciousness arise simply through an aggregation of particles. It is true that he speaks often enough as though there were not the slightest difficulty in that regard—for instance, when he tells us that "it is the nature of Matter, when raised corpuscularly to a very high degree of complexity, to become centred and interiorized—that is to say, to endow itself with consciousness."[9] At first there are only particles; and then, once a sufficiently high degree of complexity has somehow been attained, consciousness appears upon the scene: that is clearly the message. Yet at other times we are given to understand that the question is not really quite so simple, and that in fact one needs to assume that consciousness is there from the start. "We are logically forced to assume," so reads one of these passages, "the existence in rudimentary form (in a microscopic, i.e., an infinitely diffuse state) of some sort of psyche in every corpuscle, even in those (the mega-molecules and below) whose complexity is of such a low or modest order as to render it (the psyche) imperceptible."[10]

But why? If it be the nature of Matter "to endow itself with consciousness" when it has attained "a very high degree of complexity," why must it be assumed that some kind of rudimentary consciousness exists even in the simplest of corpuscles? One knows that numerous phenomena (shock-waves, for example) can occur only when a certain critical point or threshold has been reached; how, then, can we be certain that this is not also what happens in the case of consciousness? Why must one assume that consciousness (or "some sort of psyche") exists in an electron, say, or in a gas? And if consciousness can be adequately accounted for by "a very high degree of corpuscular complexity," why is one "logically forced" to assume that it exists where there is no such organized complexity at all? In fact, how can the "specific effect" exist in the absence of its

9. FM, p. 226.
10. PM, pp. 301–302.

specific cause? It would seem that if one is indeed compelled to postulate "some sort of psyche in every corpuscle," it could only be for the reason that the psyche is *not*, after all, conceivable as the specific effect of organized complexities.

But be that as it may, it is clear, in any case, that there are no empirical grounds whatever substantiating the existence of an "infinitely diffuse" psyche or consciousness. Now, it happens that Teilhard himself is aware of this difficulty, and has tried to account for it: one assumes the existence of a rudimentary psyche, imperceptible though it be, he tells us, "just as the physicist assumes and can calculate those changes of mass (utterly imperceptible to direct observation) occasioned by slow movement."[11] The reference, of course, is to the relativistic stipulation that the mass of a particle increases with its velocity by a factor which remains exceedingly close to 1 so long as the velocity is small compared to the speed of light. But, in the first place, if "direct observation" means measurement, it is not true that the changes of mass occasioned by slow motion are "utterly imperceptible": they remain in principle "perceptible" right down to the point where quantum effects supervene. And at that point, not only does measurement become impossible, but the relativistic theory itself breaks down: we are then in a range where no one knows precisely what in fact is going on. Moreover, the effect in question (where it does apply) is predicted on the strength of a mathematical theory which can in principle be tested and verified. And if indeed it should happen that some of its predictions cannot be tested directly, this obviously does not bestow scientific sanction upon every gratuitous assumption of an unverifiable kind! We need not belabor the point: rhetoric aside, there is absolutely no scientific basis for Teilhard's contention that "some sort of psyche" exists in every corpuscle.

It is questionable, in fact, whether the notion of an "infinitely diffuse" psyche or consciousness actually makes sense. Consciousness, as we have noted, is inherently a power, a certain *pouvoir*, and as such it is an aptitude for acts of some specific kind. To speak of an "infinitely diffuse" consciousness is therefore very much like allud-

---

11. Ibid., p. 302.

ing to a vector which has neither a magnitude nor a direction: clearly, such a "vector" simply does not exist.

But why does Teilhard postulate so vehemently? Why does he repeatedly make dogmatic pronouncements which prove to be baseless and at times contradictory? What is he driving at: what is the actual point of this curious dialectic? It is not easy, of course, to answer these questions, which reach beyond the persona, the visible man. It may not, however, be far amiss to suggest that Teilhard's real concern—and the one thing he is careful never to contravene—is precisely *the abolition of the traditional dualism*. As we have already noted, such is in fact the hidden thrust of that so-called Law upon which his entire theory is based. One has the impression, at times, that it hardly matters to him whether consciousness springs into existence suddenly, as it were, as a direct result of some fortuitous conjunction of particles, or preexists in some rudimentary and undetectable form. He seems, in any case, to vacillate in that regard, and leans now to this and now to that side. Yet in either case he denies the traditional dualism: that is just the point. He denies the traditional doctrine when he aligns himself, as he does often enough, with those who delight in explaining everything by way of physics and chemistry, and he does so again when he proclaims some kind of pan-psychism or monism based upon a so-called spirit-matter. It seems that just about anything goes, with one notable exception, which happens to be precisely the metaphysical doctrine Christianity has made its own. This is what Teilhard consistently denies, and for very good reason: for it is ultimately the one teaching which irremediably proscribes the postulate of radical evolutionism.

But whereas Teilhard persistently attacks the very foundations of the traditional doctrine, he freely avails himself of Christian tenets which he believes can be incorporated into his theory. An especially noteworthy case in point is the doctrine of human immortality, which he wishes to appropriate in its most uncompromisingly Christian form. It is not just some universal principle or substance—

some amorphous ground—that supposedly survives the dissolution of the human compound, but the human person itself: sometimes, at least, Teilhard gives us to understand, in conformity with Christian belief, that it is truly Peter and Paul who survive. But the question remains whether this orthodox teaching can in truth be maintained within the framework of Teilhard's unorthodox thought: does it even make sense?

"By death, in the animal," Teilhard writes, "the radial [energy] is reabsorbed in the tangential, while in man it escapes and is liberated from it."[12] Now "radial energy," the mysterious factor which, according to Teilhard's theory, draws the organism towards higher levels of complexity and consciousness, is obviously but another term for the soul. What Teilhard is saying, therefore, is that at death the human soul detaches itself from the body and continues to exist in another state: it is basically just the orthodox tenet, decked out in scientific-sounding terms. Where he differs from the Christian teaching, on the other hand, is in what he has to say regarding the *origin* of this soul: for whereas Christianity insists that the human soul is created by God *ex nihilo*, Teilhard argues that it has somehow evolved out of the primordial "stuff of the universe." According to his theory, the soul comes into existence gradually, through the evolutive process of complexification, until it attains the level of man, at which point it turns reflective. At the moment of death it then passes through a second critical point and becomes "detached" from the body. It then rises "like a trembling haze that vanishes,"[13] Teilhard tells us poetically. "All around us, one by one, like a continual exhalation, 'souls' break away, carrying upwards their incommunicable load of consciousness."[14]

Now this may conceivably be true in some appropriate sense; but how does he know? Is he still speaking from that "phenomenal point of view, to which I systematically confine myself"? Has he actually perceived that "trembling haze"? We are spared such a claim. These things are said from the same ostensibly sober and scientific stand-

---

12. Ibid., p. 272.
13. HM, p. 190.
14. PM, p. 272.

point from which he speaks about organic evolution, hominization and the rest; but whereas Teilhard's notion of soul in the sense of "radial energy" might conceivably have a certain scientific sense when taken in a *bona fide* biological context, where the elusive factor could perhaps be investigated by way of its observable effects, it is clear that by the time Teilhard speaks of "invisible exhalations" he has altogether departed from the domain of scientific discourse. The question therefore remains: how does he know? Is this simply a conjecture, a beautiful and perhaps most desirable notion with which to cap his system? Or does he, perchance, base himself on the authority of Christian tradition, and so, ultimately, on Revelation? But in that case, why would he disregard what Christianity has to say on other matters? If the ground of Revelation is solid enough to stand upon when it comes to the soul's end, why not also in regard to its origin?

We need not however concern ourselves with these rather hypothetical questions. Inasmuch as Teilhard has chosen to take his stand on scientific ground, his claims should first of all be judged on that basis. How then, let us ask, does the Teilhardian theory of human immortality stand scientifically? And the answer can only be: very badly indeed. It could even be argued, quite cogently, that the very notion contradicts his fundamental postulate, namely, the so-called Law of Complexity/Consciousness. For if it be the case that somatic complexity begets consciousness as its "specific effect," there must obviously be some nexus, some necessary connection, between these two aspects or components of the living organism: the corporeal, that is, and the psychic. Then death supervenes, and in an instant that "necessary connection" is broken; thought or consciousness, which just a moment ago rested squarely on the support of brain function as its specific effect, breaks away and floats off, as it were, into outer space. Perhaps it was a blow to the head or a bullet that shattered the physical instrument of thought. Every tenet in Teilhard's system would lead one then to conclude that this must be the end of thought, the end of consciousness, the end of that "person" who has emerged precisely through the formation of that physical instrument. Yet suddenly, and for no discernible reason, the instrument is no longer needed: the brain ceases to be necessary simply by virtue of being destroyed.

It is to be noted that the traditional teaching avoids this absurdity. So long as the soul is not created or brought into existence through an aggregation of corpuscles, it makes sense to maintain that it continues to exist when these corpuscles are again dispersed. Admittedly the soul may be impoverished or restricted in certain respects through the loss of its bodily instrument, which after all was created for its use. Let there be no doubt in that regard: Christianity insists that the body does serve a purpose: it is not just a "prison" of the soul as some Platonists maintained. But though it has its use, it is not the cause of the soul, not the source of consciousness, or of the human intellect. And that is the crucial point of difference between the Christian and the Teilhardian anthropology.

According to the traditional teaching—Christian or non-Christian as a matter of fact—the soul has an existence of its own and a certain autonomy right from the start. It does not suddenly acquire that autonomy at the moment of death as Teilhard maintains—as if a spiritual agent could be created simply by the destruction of a physical instrument. Now it is true that in its embodied state the soul sees through the eye, hears through the ear, speaks through the tongue, and thinks through the brain. But *it does not know through the brain*! And that is the essential point.[15] The traditional anthropologies, without exception, maintain that the soul knows, not through a corporeal instrument, but by virtue of an incorporeal faculty called by various names, ranging from the Vedantic *buddhi* to the Scholastic *intellectus*. As St. Thomas explains: "The intellect is a faculty of the soul, and the soul is the form of the body; but the power that is called the intellect is not the actualization of any bodily organ, because the activity of the body has nothing in common with the activity of the intellect."[16] In a word, the human soul *knows* by way of a faculty which has "nothing in common" with bodily function; and that is precisely the reason why that soul is able, not only to exist after death, but to be cognizant "intellectu-

---

15. For and in-depth discussion of this crucial question I refer to my chapter on "Neurons and Mind" in *Science and Myth*, op. cit.

16. *Opusculum, De unitate intellectus contra Averroistas*, iii; quoted in J. Rickaby, *Of God and His Creatures* (Westmister, MD: Carrol Press, 1950), p. 127.

ally" in its postmortem state. As Aristotle had already surmised: "There is nothing to prevent some parts of the soul being separable from the body, because they are actualizations of nothing corporeal."[17] Intellectual life need not end with death: we may yet *know*, and perhaps even to an incomparably higher degree than in our embodied state, for only those acts that require a corporeal instrument are abrogated at death. And so, too, only those "parts" of the soul which depend upon physical organs for their actualization are affected. From a trichotomous point of view one could say that the *anima* or *psyche*, properly so called, becomes "absorbed" after death in the purely spiritual part of man, the *pneuma* (which includes the intellect). The *pneuma* as such, on the other hand, is not affected.

Though it may be based on Revelation or upon the spiritual experiences of mystics, the traditional teaching proves nonetheless to be exceedingly logical: the more closely it is probed, the more the parts are perceived to fit together with a marvelous—one might almost say "mathematical"—precision. Consider its claim of human immortality. We have noted that the act of knowing demands, not an extreme complexity as Teilhard would have us believe, but just the opposite: for only what is in fact supra-spatial and uncompounded could transform a spatially organized multiplicity into an object of the cognitive act, which is perforce *one* thing. But by this very token one likewise discovers that the intellect as such is immortal, that it survives, for it is self-evident that what is perfectly simple cannot be destroyed: only complexity is vulnerable, only compounds can be dissolved. And so we find that the tenet of human immortality is altogether consonant with the traditional understanding of the intellectual act. In point of fact, it accords perfectly with every facet of the integral doctrine.

The case is very different, however, when this dogma has been cut off from the traditional teaching at large and grafted into the evolutionist theory of Teilhard de Chardin. As we have already noted, the resultant theory is plainly incongruous. Under the stipulations of radical evolutionism the Christian tenet of immortality becomes not only gratuitous but indeed absurd. Like the farfetched invention

17. *De Anima*, II. i. 12.

of an inept playwright, Teilhard's "discarnate soul" can finally be nothing more than a *deus ex machina* called forth, as if by fiat, to save a sinking plot. But it does not in fact save the plot, but only adds confusion. It would seem that when Teilhard speaks elsewhere of "the principle of coherence" as the prime criterion of truth,[18] he has first of all condemned his own theory.

Teilhard has always made it a point to present himself, first and foremost, as a man of science. One might be tempted to think that the claim could be tendentious: after all, Carl Jung was no doubt right when he remarked—with reference to Sigmund Freud—that "today the voice of one crying in the wilderness must strike a scientific tone if the ear of the multitude is to be reached."[19] To be sure, neither as a philosopher, nor as a theologian, could Teilhard have had any comparable impact upon "the multitude." He needed the prestige and charisma of the scientist to captivate his audience, and he knew very well how to capitalize upon these assets. And yet nothing we know contradicts the notion that Teilhard did have unbounded admiration for science, combined with an equally high estimate regarding the scientific worth of his own ideas. But be that as it may, what we do know for certain is that Teilhard's far-flung speculations concerning organic evolution, so-called hominization, noögenesis, cosmic convergence and so forth, are not in fact scientific: it is a long way from fossils and the skeletal remains of a conjectured Sinanthropus to Point Omega![20]

---

18. See, for instance: HE, p. 94; CE, p. 130n; and FM, p. 222.
19. *The Collected Works* (NY: Pantheon, 1971), vol. 15, p. 38.
20. It is interesting that Teilhard himself has admitted this, and has in fact repudiated his scientific pretensions in one of his letters, wherein he writes: "I sense how in itself the exploration of the Earth can bring no light, and does not enable us to find any solution to the most fundamental questions of life. I have the impression of moving around an immense problem without being able to penetrate into it. Moreover, as I also observe, this problem appears to wax before my eyes, and I see that its solution is to be sought nowhere but in a 'faith' which goes beyond all experience. It is necessary to break through and pass beyond appearances." (*Lettres de Voyage, 1923–1955*, Edition Grasset, p. 31).

One must also remember that Teilhard's renown as a kind of universal scientist has been trumpeted mainly in more or less theological circles. Certainly, in his own field, Teilhard did command the respect of his fellow scientists, as is evidenced by the fact that he was made a member of the prestigious *Academie des Sciences*, and a director in the *Centre National de la Recherche Scientifique*. So far as his wide-ranging theories are concerned, on the other hand, the reaction of scientists has been mixed and generally guarded. On the enthusiastic end of the spectrum, mention should be made of Sir Julian Huxley, who introduced *The Phenomenon of Man* to the Anglophone world with a warm and at times exuberant commendation. Sir Julian goes so far as to say, with specific reference to the Law of Complexity, that although "this view admittedly involves speculation of great intellectual boldness," it is "extrapolated from a massive array of fact, and is disciplined by logic. It is, if you like, visionary: but it is the product of a comprehensive and coherent vision."[21] Yet despite this positive judgment—with which we strongly disagree!—the celebrated evolutionist stops short of endorsing Teilhard's contention that *The Phenomenon of Man* is in fact "purely and simply a scientific treatise"[22]: whatever else the Teilhardian doctrine may be, a "scientific treatise" it most definitely is not.

As concerns the merits of that doctrine, scientists have held vastly different views. Among the harshest critics one should above all mention the Nobel laureate Peter Medawar, whose review of the aforesaid treatise amounts to a blanket condemnation of the Teilhardian theory as a whole. He first of all dispels the stereotyped image of the "great scientist," which by that time had come to surround the figure of Teilhard de Chardin, in the eyes of his followers, like a nimbus. "Teilhard practiced an intellectually unexacting kind of science," Sir Peter tells us, "in which he achieved a moderate proficiency. He has no grasp of what makes a logical argument or what makes for proof. He does not even preserve the common decencies of scientific writing, though his book is professedly a scientific

---

21. PM, p. 16.
22. Ibid., p. 29.

treatise."[23] Besides objecting to a generally extravagant use of language, involving such literary excesses as "nothing-buttery," Medawar charges that "Teilhard habitually and systematically cheats with words."[24] What he means thereby is that Teilhard "uses in metaphor words like energy, tension, force, impetus, and dimension *as if* they retained the weight and thrust of their special scientific usages." And this is presumably the prime offense against "the common decencies of scientific writing" which Medawar holds against the French Jesuit. Yet there are still other improprieties, he charges, not the least of which is self-contradiction; as Sir Peter points out by way of example: having stated that "complexity increases in geometrical progression as we pass from the protozoon higher and higher up the scale of the metazoa," Teilhard goes on to inform the reader that "the nascent cellular world shows itself to be already infinitely complex." The Nobel laureate finds this disturbing, as indeed it is.

Yet when Medawar goes on to express his overall appraisal of the Teilhardian opus by saying that "the greater part of it, I shall show, is nonsense, tricked out by a variety of tedious metaphysical conceits,"[25] many have felt that he has gone too far. And to be sure, it is not at all clear what precisely he means by this indictment, and on what basis it has been made. After all, there are those—Nobel laureates included—who perceive every facet of metaphysical doctrine as "a conceit." But this is a question which hardly concerns us: if Medawar may have been too harsh in his overall judgment, the fact remains that he would certainly have recognized *scientific* merit if he had encountered such.

And there is yet something else that needs very much to be pointed out: in what appears to be his central criticism—i.e., that "Teilhard habitually and systematically cheats with words"— Medawar has no doubt hit the nail on the head. The word "cheat" is of course very strong, and should not be interpreted in its literal sense, which is "to deceive by trickery, to defraud, to swindle"; it should rather be understood in its milder sense, which is "to fool"

---

23. *Mind*, vol. 70 (1961), p. 105.
24. Ibid., p. 101.
25. Ibid., p. 99.

or "to beguile," without the implication of ill intent. What alone concerns us, in any case, is the fact that a certain systematic misuse of language does run through the Teilhardian opus, and does tend "to fool and beguile" the reader, as the history of the post-Vatican II era proves. Medawar has in fact hit the mark when he observes that "it is the style that creates the illusion of content." Precisely! And the single most effective and most frequently applied "device" is undoubtedly the misuse of metaphor.

This is what is going on, for example, when consciousness is said to be "a dimension, something with mass, something corpuscular and particulate which can exist in various degrees of concentration, being sometimes infinitely diffuse."[26] And the same thing, to be sure, is happening when Teilhard speaks of discarnate souls in terms of vapors, exhalations, or bubbles, not to mention "a trembling haze." Whatever merit such metaphors may have from the standpoint of poetry, their scientific worth is nil. Yet in countless passages vital to his doctrine, Teilhard is speaking in terms of metaphors, which may sound poetic or have a scientific ring, but are always false the moment one forgets that they are—not rigorous conceptions, be they scientific or philosophical—but indeed metaphors. The point is that Teilhard does seem to forget this, not just here and there, but systematically: for example, in *The Phenomenon of Man*, when he tells us that he is speaking from a strictly "phenomenal point of view," and that the book is to be read "purely and simply as a scientific treatise." The "trembling haze" ceases thus to be a poetic figure of speech, and becomes forthwith a scientific phenomenon. And this is of course precisely what thousands—including unfortunately theologians of high rank—have come to believe. To the admiring multitude the Teilhardian pronouncements have become oracles of Science; and the more flagrantly far-fetched, the more earth-shaking do these pronouncements appear.

Even so the fact remains: take away the metaphors and nothing is left. And this is just what Medawar means when he charges that "it is the style that creates the illusion of content." Teilhard's metaphors are not simply an embellishment, or a means to explain some diffi-

---

26. Ibid., p. 101.

cult scientific ideas to a nontechnical audience: in essence they *are* the theory. That is what even Sir Julian has failed to grasp when he speaks, quite innocently, of Teilhard's "genius for fruitful analogy."[27] Analogy with what? With another analogy, perhaps? What is lacking in Teilhard's doctrine are scientific definitions, *bona fide* scientific conceptions, which can then perhaps be explained or illustrated in terms of "fruitful analogies."

27. PM, p. 20.

4

IN SEARCH OF CREATIVE UNION

THERE ARE THOSE who believe that Teilhard has laid the foundations of a new metaphysics based upon scientific facts. As Jean Danielou has put it: "He translates the scientific categories into metaphysical categories."[1] Or still more emphatically: "He builds a metaphysics as an extension of the science of his day."[2] Now this presumed metaphysics is none other than the celebrated Teilhardian doctrine of "creative union"; it is only that Teilhard himself appears to be less certain than his Jesuit confrère that the theory is in fact metaphysical. "Creative union is not exactly a metaphysical doctrine," he notes. "It is rather a sort of empirical and pragmatic explanation of the universe, conceived in my mind from the need to reconcile in a solidly coherent system scientific views on evolution (accepted as, in their essence, definitely established) with the innate urge that has impelled me to look for the Divine not in a cleavage with the physical world but through matter, and, in some sort of way, in union with matter."[3] What this means, in plain terms, is that Teilhard was led to his notion of creative union in an effort to reconcile the rudiments of Darwinism with his own pantheistic propensities. It is not clear, of course, how a synthesis based upon Darwinist and pantheistic premises could conceivably result in "a sort of empirical and pragmatic explanation of the universe." But be

---

1. "Signification de Teilhard de Chardin," *Études*, vol. 312 (1962), p. 147.
2. Ibid.
3. SC, p. 44.

that as it may, let us try, in any case, to discover what that "explanation of the universe" turns out to be.

Now it happens that Teilhard alludes to the notion of "creative union" innumerable times; and yet there are only a few pages, here and there, where the matter is dealt with explicitly. One of these expositions—which happens to be particularly revealing from a scientific point of view—is to be found in an essay entitled "Human Energy." Written seventeen years following an introductory note, "Creative Transformation," it can be taken as representing a relatively mature stage in the unfolding of Teilhard's thought on that issue. It behooves us, therefore, to examine the essay with some care.

Teilhard begins with a statement underlining the empirical nature of what is to follow: "A principle of universal value appears to emerge from our outer and inner experience of the world, which might be called the 'principle of the conservation of personality.'"[4] And what is that principle? "At *a first stage*," he goes on to say, "the law of conservation of personality only states that the rise of spirit in the universe is an *irreversible* phenomenon"[5]; or more succinctly still: "Conservation (without regression) of the highest stage of personalization acquired at each moment by life in the world."[6]

But there is more. "At *a second stage*, the principle of conservation of personality suggests that a *certain amount* of energy, in the impersonal state, is engaged in the evolution of the universe, and that it is destined to be transmuted entirely into a personal state at the end of the transformation (the quality of this 'personal end-product' being moreover a function of the quantity of 'impersonal' material engaged at the beginning of the process)."[7] And this explanation is followed by what is presumably a concise statement of the principle itself: "Conservation (without loss), in the course of the spiritualization of the universe, of an undefined amount of power or cosmic 'stuff.'"[8]

---

4. HE, p. 160.
5. Ibid.
6. Ibid., p. 161.
7. Ibid.
8. Ibid.

Obviously this is all very vague. But be that as it may, we are told, in any case, that "under this absolute, *quantitative form*, the law of conservation of personality is not directly capable of demonstration, perhaps because it refutes our formal knowledge that we are able to measure the world by 'cubing' it, or perhaps because we do not yet see how to express the coefficient of transformation from impersonality to personality."[9] Meanwhile, however, a far simpler and more prosaic reason for the stated unverifiability will perhaps have occurred to the perceptive reader: could it not be that the presumed principle, thus formulated, has in fact no scientific content at all?

What mainly concerns us at this point, however, is to elucidate whatever *philosophical* sense there might be in these Teilhardian speculations. We are searching, after all, for the foundations of that "evolutionist ontology" which has impressed so many theologians of the new school.

Unverifiable, then, though it be, Teilhard informs us that the principle "has nevertheless a use: it states that the spiritualization taking place in the cosmos must be understood as a *change of physical state* in the course of which a certain constant is preserved throughout the metamorphosis."[10] For a moment one has the impression that the picture is beginning to come into focus; but then the realization dawns that we have not the slightest idea what that "certain constant" might be. Could it be the amount of physical energy? Or is it something else? And why are we not told the very thing, in the absence of which we have in fact been told nothing at all?

"Understood in this way," Teilhard continues, "the conservation of personality in no way implies (quite the contrary) an 'ontological' identity between the unconscious and the self-conscious. Although subjected to a 'quantic' law, personalization remains in effect essentially an evolutionary transformation, that is to say, continually the generator of something entirely new. 'So much matter is needed for so much spirit; so much multiplicity for so much unity. Nothing is lost, yet everything is created.' This is all that is affirmed."[11] But

9. Ibid.
10. Ibid.
11. Ibid., pp. 161–162.

again the matter is not quite so simple: a crucial ambiguity—which has been plaguing us all along—needs to be resolved. There are namely two conceptual possibilities: either matter (or physical energy, which amounts to the same) is actually transformed into so-called "spirit," or it is not. Now, which shall it be?

In numerous places Teilhard appears to have opted for the first alternative. One has this impression, for example, when he speaks of energy "destined to be transmuted entirely into a personal state," and of a corresponding "coefficient of transformation." And this is in fact what generally seems, at least, to be Teilhard's position when he speaks of a metamorphosis, or a "change of state," and appears also to be expressed as clearly as one could wish in the oft-quoted Teilhardian formula: "All that exists is matter becoming spirit."

It is interesting to observe, moreover, that under these auspices the rise of consciousness would presumably be accompanied by a certain diminution of physical energy. And this—at long last!—is something that could in principle be measured and thus put to the test scientifically. Nor would one have to know "how to express the coefficient of transformation from impersonality to personality" as Teilhard suggests: the mass or energy defect associated with an irreversible "psychogenesis" is after all an observable. If "so much energy" is consumed in the production of "so much consciousness"—whatever this may mean!—then there is exactly "so much" less energy in the system after the postulated metamorphosis has occurred. And let us add that if this were found to be the case, Teilhard's "principle of psychogenesis" would evidently constitute one of the greatest discoveries in the history of science.

But needless to say, there is not the slightest reason to suppose that such a mass or energy defect is to be found; and as we have seen, Teilhard himself has made it a point to inform us that his principle, in its so-called quantitative form, is not subject to verification. It therefore appears that we should perhaps adopt the second line of interpretation: that matter is not, after all, transformed into spirit. And this is presumably what Teilhard himself is suggesting when he declares that "the conservation of personality in no way implies (quite to the contrary) an 'ontological' identity between the unconscious and the self-conscious," and adds that the transformation is "continually the generator of something entirely new." Energy, in the sense of the physicist, would then be strictly conserved, and the so-called metamorphosis would reduce to "a change in physical state," characterized by a certain complexification. Consciousness or spirit, in that case, would then be something entirely new, something which does *not* come forth out of matter by way of a *bona fide* metamorphosis. Is this, perhaps, what Teilhard means when he says that "nothing is lost, yet everything is created"? But that would imply that he has opted for the second alternative: the idea that spirit is something utterly new.

But then, under these auspices, what is to be made of the following statement, which Teilhard has put at the end of the entire discourse, and thus presumably as its conclusion: "In a universe where spirit is considered *at the same time* as matter, the principle of the conservation of the personality appears as the most general and satisfactory expression of the invariance of the cosmos first suspected and sought by physics on the side of the conservation of energy."[12] Here Teilhard seems to be leaning, once again, towards the first alternative: for what else could this statement mean than that matter and spirit are two forms or aspects of a single energy, a single "power or cosmic 'stuff'" as Teilhard has put it earlier, which is the one thing that is conserved? And this would mean that energy, in its strictly physical manifestation, would be conserved only to the extent that psychogenesis is not taking place, and that there *is*, after all, an 'ontological' identity between matter and spirit—or "between the unconscious and the self-conscious"—which indeed brings us back to the first alternative: the notion that physical energy is in fact transformed into "spirit."

If we return to the second interpretation, on the other hand, Teilhard is saying that although a certain aggregation of material particles is required to effect the emergence of spirit, yet according to his principle of conservation (at least in the case of man) that "quantum of spirit" continues to exist even after the aggregate in question has been dissolved. But in that case his so-called "principle of the conservation of personality" is not in fact a law of conservation, but

12. Ibid., p. 162.

the very opposite: for to say that something comes out of nothing and remains is to deny the existence of an invariant, i.e., of something that is conserved. Teilhard's theory reduces then, not to a monism—a doctrine which speaks of a single "cosmic stuff" within which "changes of state" occur—but a dualism involving two irreducible principles, one of which requires the other in order to emerge.

But perhaps we have still not fathomed the true meaning of Teilhard's less-than-precise affirmations. Could it be that the emergent spirit is neither matter transformed, nor something that emerges quasi *ex nihilo* in the aforementioned sense, but is rather to be conceived as the manifestation of something immaterial that was there all along? But in that case the emergence of spirit is neither a transformation nor the creation of something new; and this means that Teilhard would have contradicted himself, not once, but twice. Yet there is nothing to suggest that he himself was in any way disturbed by these ambiguities and apparent contradictions; one has rather the impression that he actually wants to have it both ways. Teilhard seems blithely to ride the razor-edge of a logical alternative, leaning now to one side, now to the other, without ever committing himself either way, or even evincing a need to do so.

What, then, are we to make of this apparent confusion? Is it perhaps a mark of unusual profundity, as some have tended to assume? Or an indication, conceivably, that the law of the excluded middle has itself been superseded in a Darwinist universe? But fortunately we need not ponder these questions; for Teilhard himself has made it clear that his theory of creative union has been advanced, in a perfectly sober and scientific spirit, as "a kind of empirical and pragmatic explanation of the universe," and is supposed to constitute "a solidly coherent system." But as we have seen, it does not remotely meet either of these requirements.

Yet it is easy enough to understand what Teilhard was attempting to do. It is evident that he was fascinated with the idea of "creative union," and charmed by such formulas as *"Deus creat uniendo"* or *"creari est uniri."* As Henri de Lubac has put it: "Such axioms appealed to him and led him to dream of constructing a metaphysics, his own metaphysics, which would be a 'metaphysics of union.'"[13] But it must also be remembered that this close friend and ardent admirer of Teilhard de Chardin concludes his discussion on the subject with the observation: "We must, nevertheless, admit that he did not achieve a perfectly clear and coherent formulation of his thought."[14] In plain words: Teilhard's dream of constructing a metaphysics of his own was never realized.

Yet ill-defined though it be, inasmuch as the notion of creative union is obviously central to the Teilhardian doctrine at large, it behooves us to consider the use to which that protean notion is put. There are, in particular, two sections in the well-known essay, *Mon Universe*, which need to be closely scrutinized.

"Creative union," so the exposition begins, "is the theory that accepts this proposition: in the present evolutionary phase of the cosmos (the only phase known to us), everything happens as though the One were formed by successive unifications of the Multiple,"[15] And as Teilhard goes on to explain, this does *not* mean "that the One is composed of the multiple, i.e., that it is born from the fusion in itself of the elements it associates (for in that case either it would not be something created—something completely new—or the terms of the Multiple would be progressively decreasing, which contradicts our experience),"[16] What is being asserted is simply "that the One appears in the wake of the Multiple, dominating the Multiple, since its essential and formal act is to unite."[17]

Let us note that this is the kind of affirmation that has reassured many an orthodox reader, and tempted some interpreters to view Teilhard as a virtual Thomist. But the point is that one must read on.

"At the lower limit of things," we are informed presently, the so-called law of recurrence "discloses an immense plurality—complete

---

13. *The Religion of Teilhard de Chardin* (NY: Desclee, 1967), p. 196.
14. Op. cit., p. 200.
15. SC, p. 45.
16. Ibid.
17. Ibid.

diversity combined with total disunity."[18] It seems to be a settled conviction with Teilhard de Chardin that everything begins in multiplicity and converges towards an ever-greater unity. And yet even the most elementary observations disclose just the opposite! The fertilized ovum, for example, which looks like a tiny globule, divides and subdivides, creating a spherical immensity of cells. Then the blastosphere invaginates, and the cells begin to specialize, thus giving rise to a multiplicity of layers, tissues, and organs. The entire movement appears to be directed from unity to multiplicity. And what is death, after all, but the final victory of multiplicity on the organic plane.

It may not be without interest to point out that one meets a similar spectacle in other domains: in the sphere of art or the world of thought, for instance. Here too one can observe what appears to be a pre-existent unity, progressively unfolding a multiplicity from out of itself. Every writer, every creative scientist, every thinker has witnessed this process; it is happening everywhere. We have all experienced it: an idea is born in our mind—at one stroke, as it were; and under the influence of a certain brooding, it swells and unfolds into a multiplicity of some kind, which again, by stages and degrees, complexifies further, till it attains its full-blown state. And let us not forget Mozart's famous testimony to the effect that an entire symphony could first present itself in the form of a single musical idea, conceived all at once, in a flash of inspiration.

Yet Teilhard seems to be convinced—no one knows why—that things invariably move in the opposite direction: from multiplicity to unity. This is the first premise, in any case; and the second is stranger still: not only do all things begin in multiplicity, but it is unity that unites them. We are told, for example, that "in the first stages in which it becomes conceivable to us, the world has already been for a long time at the mercy of a multitude of elementary souls that fight for its dust in order that, by unifying it, they may exist."[19] But the difficulty with this presumed explanation is that it is not in fact conceivable. Even a confirmed Darwinist, we suspect, might

---

18. Ibid., p. 47.
19. Ibid.

find it hard to understand how a "complete diversity, combined with total disunity," could give birth to "elementary souls," which exist by virtue of material aggregates *formed under their influence*! How, in other words, can "elementary souls" exert an influence *before* they have been brought into existence? With this kind of logic, it seems, one can prove just about anything.

But let us go on to the third premise: "Only in man, so far as we know, does spirit so perfectly unite around itself the universality of the universe that, in spite of the momentary dissociation of its organic foundation, nothing can any longer destroy the 'vortex' of operation and consciousness of which it is the subsisting centre."[20] Now this, too, is very strange. How can a "vortex of operation" continue to exist when it is no longer operating on anything? Does not Teilhard himself, as a matter of fact, inform us three pages later that spirit "does not 'hold together', except by 'causing to hold together'"?[21]

But perhaps we are still thinking in now antiquated "immobilist" terms. "In the system of creative union," Teilhard goes on to explain, "it becomes impossible to continue crudely to contrast Spirit and matter. For those who have understood the law of 'spiritualization by union,' there are no longer two compartments in the universe, the spiritual and the physical: there are only *two directions* along one and the same road (the direction of pernicious pluralization, and that of beneficial unification)."[22] But then, what has happened to "souls": those "elementary souls," for example, which supposedly are "fighting" for the dust of the world? Is it a direction, a vector that is fighting another vector: the future, perhaps, fighting the past? And in the case of the human compound, if such it may be called, how are we to understand what happens at the moment of death? Until just yesterday all the world thought that here, at this fateful juncture, soul and body part company, and even Teilhard speaks of a "vortex of operation" which somehow detaches itself from the material aggregate. Is it a vector, then, that dissociates itself

20. Ibid., p. 47.
21. Ibid., p. 50.
22. Ibid., p. 51.

from another vector? But even so there are *two*: two logical "compartments," just as before, when we still persisted "crudely to contrast Spirit and matter." Or are we to say, perhaps, that body and soul are one and the same thing until death supervenes, at which point the one becomes two? We find it hard to agree with Teilhard when he says, with reference to his new theory: "Thus those innumerable difficulties vanish...."[23]

What Teilhard would like to say, but can't, is that there is Evolution and nothing else. His position is somewhat reminiscent of Heraclitus (and of certain Buddhist philosophers): there are no "things," no substances or natures in the universe, but only movement, only change, only a perpetual *genesis* or becoming. All that exists is flux. Strictly speaking, there is no cosmos even, but only a cosmogenesis. And that is Evolution.

But let us note, first of all, that this is not what Heraclitus actually taught. Yes, "Everything flows," and this world, with its seemingly solid parts, is in truth "an ever-living fire"; but it is a fire "kindled in measure and quenched in measure": that is the crucial point. True to the genius of his race, Heraclitus perceived that the flux does not stand alone, but is subject to measure, bounded by a law. The world moves, but the law remains fixed. There is flux, but there is stasis as well. Not the seeming stasis of a stone "at rest" (which is relative, and in a way illusory), but a stasis that is transcendent: a *logical* stasis, in the truly Greek sense of that term. In the final count, flux and stasis imply one another; and this entails that the cosmos, the psycho-physical reality, partakes somewhat of both. There is that which moves, and that which remains unmoved; in the language of science, there are variables, and there are invariants. And what counts are in fact the invariants; this is just what physics is about: it is a search in quest of invariants. What matters to the scientist is not the "ever-living fire" as such, but the "measures" in which that fire has been "kindled and quenched," to put it in the expressive terms of our Greek philosopher.

23. Ibid.

And this brings us to the crucial point: these measures derive "from above," that is to say, from a *spiritual* plane in the authentic sense of that much-abused term. As we have noted in a previous chapter, what renders the world intelligible is its spiritual content.[24]

Now it is interesting that Teilhard himself seems at times to be saying exactly the same thing. In *Mon Universe*, for example, he writes that "nothing in the universe is intelligible, living and consistent except through an element of synthesis, in other words a spirit, or from on high."[25] And again: "'All consistence comes from Spirit.' In that we have the very definition of creative union."[26] Once again we seem to find ourselves, momentarily, on orthodox ground. And how well Teilhard has put it when he goes on to say: "The materialist philosopher, therefore, who looks at a lower level than soul for the solid principle of the universe, grasps no more than dust that slips between his fingers"![27]

What, then, are we to make of these seemingly orthodox affirmations? Now, to begin with, we need to recall an essential point: as we have noted in Chapter 2, Teilhard has in effect rotated the *axis mundi* of the perennial doctrine through ninety degrees, so as to make it coincide with the stipulated vector of Evolution. The "above" has thus become the "ahead." But this means that in Teilhard's theory there *is* no "above," no Spirit in the authentic sense. There is only Evolution, only "directed flux." There is simply no room in this "one-dimensional" model for Spirit as such. There is past and future, but no metaphysical verticality, no "above": no *bona fide* Spirit. For as we have seen, Spirit, standing as it does on the side of stasis, transcends flux, *transcends evolution*.

And yet Teilhard continues to speak of Spirit as a *sine qua non*. We are told that "all consistence comes from Spirit," that "everything holds together from on high."[28] But what does this mean?

---

24. This "spiritual content" is just what metaphysics in the time-honored sense is about: from Plato's Ideas to the "forms" of the Scholastics, it has been the subject of inquiry and debate.
25. SC, p. 57.
26. Ibid., p. 49.
27. Ibid.
28. Ibid., p. 50.

What can "from on high" connote in a one-dimensional universe, a universe consisting of directed flux? Now this is precisely where the "rotation of axes" comes into play: the "above" is henceforth conceived as the "ahead." This is precisely Teilhard's master-stroke: the Darwinists and materialists at large had never thought of such a thing; they would have simply said: "All consistence comes from matter." It never occurred to them—nor perhaps to anyone prior to Teilhard de Chardin—that the cause of all consistence could lie in the "ahead": in a principle yet to be born!

There is however a difficulty with that ingenious notion: it turns out that the idea is not in fact conceivable. And is this not the reason, too, why that pseudo-doctrine has had to be so thoroughly disguised? It appears that to pass muster, the teaching demands all the vagueness, all the ambiguity, all the equivocation Teilhard could provide: it would never do simply to proclaim, in a clear voice, that "all consistence comes from the ahead." To which one might add that this very vagueness, ambiguity, and equivocation serves at the same time to bestow an impression of ungraspable profundity.

But getting back to the logical point: what Teilhard has manifestly failed to recognize—and what persistently plagues him—is the fact that there can be no movement without a corresponding stasis, no *evolution*, if you will, without something that does *not* evolve.

One might ask whether Teilhard's primary concern was actually to found a new metaphysics or was simply to impugn the old. But be that as it may, he does attack the traditional metaphysical teaching at just about every turn, which of course is hardly surprising inasmuch as that teaching stands squarely in the way of his own. Thus we are told, at sporadic intervals, that the old metaphysical categories are "immobilist," that they derive from a pre-scientific worldview, and have now become untenable in light of scientific findings, among which the discovery of Evolution, to be sure, holds pride of place. Frequently, moreover, these indictments are made in passing, as if the matter were too obvious to call for further consideration. Here and there Teilhard does however deal with these issues at some greater length, and the result proves to be invariably enlightening. Let us consider an example, taken from his article "On the Notion of Creative Transformation."

Teilhard begins his polemic by reminding us that Scholasticism admits "only two sorts of variations in being"[29]: creation and transformation, namely, conceived respectively as the production of being "out of nothing" and "from potency of the subjacent." To which we should perhaps add that this dichotomy belongs not just to Scholasticism—as if it were somehow tied to Latin Christianity and the Middle Ages—but is in fact integral to the Christian tradition as such. Yet Teilhard is critical of this doctrine. At this point it is not entirely clear whether he wishes to do away with the traditional idea of creation, or wants simply to add a third category to the list. In any case, he begins with the following point: "Besides *'creatio ex nihilo subjecti'* and *'transformatio ex potentia subjecti,'* there is room for an act *sui generis* which *makes* use of a pre-existent created being and builds it up into a *completely* new being."[30] And he goes on to explain that "this act is *really creative*, because it calls for renewed intervention on the part of the First Cause."[31] Teilhard finds it remarkable, moreover, that Scholasticism "has no word to designate this method of divine operation," seeing that it is "conceivable *in abstracto*, and is therefore entitled to a place at least in speculation," and is "probably the only one which satisfies our experience of the world."[32] And he adds: "We should, I believe, have to be blind not to see this: *In natura rerum* the two categories of movement separated by Scholasticism (*Creatio et Eductio*) are seen to be constantly fused, combined, together."[33]

Now, to begin with, it is by no means the case that "*in natura rerum*" creation and transformation "are seen to be constantly fused, combined, together," for the simple reason that creation cannot be "seen" at all. Only *what happens in time* can in any empirical

29. CE, p. 21.
30. Ibid., p. 22.
31. Ibid.
32. Ibid., pp. 22–23.
33. Ibid., p. 23.

sense be "seen" or observed, whereas creation does *not* take place in time. It cannot, because time itself comes into existence by virtue of this act: "God, therefore, in His unchangeable eternity created simultaneously all things whence times flow" to quote St. Augustine once more. This is what, above all, we need to bear in mind if we are not to misconstrue the Christian doctrine. It would thus be erroneous to think that God created the universe six thousand—or fifteen billion—years ago, as if that "beginning" were receding into the distant past with the passing of time: to think in these terms would be, once again, to conceive of creation as a temporal act. The traditional teaching, on the other hand, Christian and non-Christian alike,[34] proscribes that position, and affirms that what St. Basil terms "the instantaneous and imperceptible moment of creation"[35] is in a sense "equidistant" to all times, even as the center of a circle is equidistant to all points on the circumference. That "moment," in other words, as the veritable *nunc stans*, is contiguous to all times and places as the omnipresent Center where "every where and every when are focused"[36] as Dante declares. Creation, then, is not to be counted among events that transpire in time, nor is it in any way observable. And on both counts it differs sharply from a transformation, which obviously is something that does take place in time, and is at least conceivably observable.

Now it is true, certainly, that creation and transformation are in a sense "constantly fused," even as the center of a circle (or the pencil of its radii, if you will) is "fused" to the circumference: after all, everything in creation hinges upon the creative Act! But although creation and transformation are thus fused, they are not on that account identical, no more than the center and circumference of a circle. Scholasticism, then, and Christian tradition at large, have done well to distinguish the two conceptions.

What Teilhard is driving at, of course, when he insists that the two "are seen to be constantly fused," is that creation and transforma-

---

34. On this subject we refer again to A.K. Coomaraswamy, *Time and Eternity* (Ascona: Artibus Asiae, 1947).
35. *Hexaemeron*, 1.6.
36. *Paradiso*, 29.12.

tion cannot in truth be separated: in reality—"*in natura rerum*—the two are one and the same, which is just what the notion of "creative transformation" is meant to express. The concept supposedly unites what Scholasticism has spuriously cut asunder. And that, too, is presumably the reason why Teilhard makes it a point to marvel at the fact that Scholasticism "has no word to designate this method of divine operation," the implication being that the Thomists, blinded by their own misconceived dichotomy, could not conceive of the *tertium quid* which turns out to be expressive of the truth.

"There is not one moment when God creates, and one moment when secondary causes develop," Teilhard goes on to explain. "There is always only *one* creative action (identical with conservation) which continually raises creatures towards fuller-being by *means* of their secondary activity and their earlier advances."[37]

What has happened can now be seen: Teilhard assumes that creation takes place *in time*, and then proceeds to view the traditional doctrine in that optic. Creation ceases thus to be a single transcendent Act, and in effect breaks up into so many localized creative acts, conceived as special occurrences interrupting the normal operation of secondary causes. And quite appropriately, Teilhard finds that conception to be altogether unacceptable. The stage is now set for the presentation of the theory by which he proposes to lead us out of that ancient quandary. What needs to be done, basically, is to replace the former "discontinuous model" by a continuous one: "Creation is not a periodic intrusion of the First Cause," he tells us; it is rather "an act co-extensive with the whole duration of the universe."[38]

Now admittedly, creation is by no means "a periodic intrusion of the First Cause": of course not! And it is likewise true that, in a sense, the creative act *is* "co-extensive with the whole duration of the universe": but not in the sense of being continuously spread out over the duration of the cosmos. Quite to the contrary: the Act by which God creates is co-extensive with the duration of the universe, not because it has a duration of so many billion years, but precisely

---

37. CE, p. 23.
38. Ibid.

because it has no duration at all, which is to say that it constitutes an *atemporal* or *supratemporal* act. And as such it is in a sense omnipresent, even as the center of a circle is co-extensive with the circumference: not by being somehow spread out or multiplied, but precisely as the origin or "source" of the cosmos in its entirety, without which not a particle can exist for an instant.

What Teilhard has done is to fuse the idea of creation and of transformation into a single concept of "creative transformation" by conceiving the creative act to be both temporal and continuous, which is of course a perfectly gratuitous step, to say the least. Yet he creates the illusion of presenting an argument—and a seemingly cogent one at that—by remonstrating against the idea of "periodic intrusions of the First Cause," as if that absurd notion were indeed the gist of the Scholastic doctrine!

Is this a case of duplicity, or just plain ignorance? One cannot say for sure. We do know that Teilhard was cognizant of the traditional *omnia simul* doctrine, because apart from the fact that in those days, at least, every Jesuit *must* have known that much theology, he refers to it on other occasions.[39] But why then does Teilhard not refer to it here, in his paper on "Creative Transformation," where it is crucial? Why, in other words, does he misrepresent the traditional teaching?

The final irony, perhaps, is that the new notion makes no sense in its own right: it turns out *not* to be "conceivable *in abstracto*" as Teilhard claims. To speak of an act "which *makes use* of a pre-existent created being and builds it up into a *completely* new being" is in fact a self-contradiction: to "build up," after all, means to change, to alter in some way. But this entails the idea of continuity, of a *bona fide* transformation, which excludes the idea that the end-product is something "completely new." There is thus a logical contradiction in Teilhard's claim, which remains even when we are told that "this *act is really creative*, because it calls for renewed intervention on the part of the First Cause": even a "renewed intervention on the part of the First Cause" cannot "build up" a being into something "completely new"! It seems that God does respect logic, even if some of His creatures don't.

39. For example, in HE, p. 239.

⊕

In a later work Teilhard makes it a point to attack the Christian doctrine of creation *ex nihilo* as such. He directs his arguments in particular against "its notion of 'participated being,' a lower or secondary form of being gratuitously drawn from 'non-being' by a special act of transcendent causality, 'creatio ex nihilo.'"[40] Now it is needless to say that this notion of participated being is indeed implicit in the Christian concept of creation, and is in fact central to orthodox Christian theology. It will be interesting to see on what grounds Teilhard objects to this immemorial teaching.

"An entirely gratuitous creation, a gesture of pure benevolence, with no other object, for absolute Being, than to *share* his plenitude with a *corona* of participants of whom he has strictly no need—that could satisfy minds that had not yet awoken to the immensity of space-time, the colossal stores of energy and the unfathomable organic articulation of the phenomenal world"[41]: that is his first thrust. But why should "the immensity of space-time" or "the colossal stores of energy" contradict, or even render implausible, the idea of a gratuitous creation? Is it because the cosmos is so much larger than our "human world"? But Teilhard himself maintains that the universe in its entirety has no other *raison d'être* than to find its completion in God. There is nothing at all unreasonable, then, in the idea that God created the world, replete with all its immensities, "to share his plenitude with a *corona* of participants of whom he has strictly no need."

Following this Teilhard brings forth a second charge, which seems to have little or no connection with the immensities of space-time and energy: "We would suffer deeply," he tells us, "in the honor we pay to being, and the respect we have for God would be insulted, if all this great array, with its huge burden of toil and trouble, were no more than a sort of game whose sole aim was to make us supremely happy."[42] But here again the logic is murky: why must it

---

40. SC, p. 180.
41. Ibid., pp. 180–181.
42. Ibid., p. 181.

be supposed that the recognition of God's unbounded love and solicitude for His creatures would cause us "to suffer deeply, in the honor we pay to being," or that it would "insult the respect we have for God"? This is all very strange. In fact, it would seem that the very opposite should be the case: realizing that God has created us "out of nothing," and with no "ulterior motive," simply because He wishes to share His boundless felicity with us, we should indeed rejoice greatly, and should love and respect God all the more. But if it should happen nonetheless that we "suffer greatly" and feel "insulted," what exactly would that prove? One might well think it would only prove that we are unreasonable and utterly perverse.

Teilhard himself, however, seems not to be fully satisfied with the foregoing considerations, for he embarks immediately upon yet another line of attack: "If we could not somehow consciously feel that we cannot 'be of service to God' without God adding something to himself," he goes on to say, "that would most certainly destroy, at the heart of our freedom, the intimate driving forces of action."[43] Now this is something else: not just hurt feelings, but a kind of paralysis is at stake this time. Yet it seems questionable, in the first place, that many of us are really quite so "noble" as to spurn the offer of immortal bliss and go on strike, as it were, simply on metaphysical grounds amounting presumably to the fact that God is God: is this the best Teilhard can do?

His next point is in the same vein: "What use have we for the *selfish* happiness of *sharing* the joy of the supreme Being, when we can dream of the infinitely greater happiness of completing that joy?"[44] But is it not perfectly ridiculous to think that this creature, which according to Teilhard's beliefs has but recently learned to walk on its hind legs, should fret over conjectured limitations that might conceivably impede its happiness once it has been admitted into the very life of God? I must confess that for my part I find these suggestions absurd in the extreme. Worse than that: to my ear, at least, they have a distinctly Luciferian ring.

But let us continue. Teilhard's next argument is stated in the form

43. Ibid.
44. Ibid.

of a question: "However gratuitous we may suppose Creation 'ex nihilo' to be, is it not inevitably marked in the first place (whatever the theologians of 'participated being' may have said) by an absolute increase of unification, and therefore of unity, in the pleromised real?"[45] Now to begin with, it is unclear what Teilhard means by an "absolute increase of unification," over and above "unification" itself; and we surmise that he employs this curious expression to lend a semblance of legitimacy to the next phrase: "and therefore of unity." But be that as it may, Teilhard is saying, basically, that unification brings about an increase of unity. But this is precisely what the traditional teaching denies! What increases is not unity as such, but the *participation* of unity, or "participated unity" as one could say. Unity as such, or absolute unity, on the other hand, no less than absolute being, belongs to God alone. And to be sure, there is neither increase nor diminution in God: these are *temporal* notions, after all.

This is what traditional doctrine has to say on the issue: and be it true or false, the position is certainly not illogical. Yet it does quite obviously conflict with the evolutionist claim to the effect that unity as such is born through a process of unification—which is of course precisely the reason why Teilhard is obliged to attack the doctrine at every turn. The point, however, is that his argument carries no force: for in saying that "an absolute increase of unification" entails an increase of unity, he is doing no more than reiterate the evolutionist assumption. The argument reduces thus to a *petitio principii*: in plain words, it begs the question.

We are told next that a new ontology—a "transposition of concepts" as Teilhard calls it—is required at the present stage of human evolution "to justify the ambitions newly emerging in the heart of man."[46] Now it is strange that the justification of ambitions "emerging in the heart of man" should be deemed sufficient ground for tampering with an immemorial metaphysical doctrine. One might likewise question the recent origin of these ambitions, especially if one recalls that, long ago, there was reputedly a being who cherished a similarly grandiose design: someone, in fact, who desired,

45. Ibid.
46. Ibid.

with all his heart, to "be God." But this touches upon another question, which needs to be dealt with in a later chapter.

We come now, finally, to Teilhard's seventh and last argument: "Philosophically," he writes, "we are still living in an antiquated body of thought, governed by notions of immobility and substance."[47] The implication, of course, is that the Christian doctrine of participated being hinges supposedly upon what Teilhard calls "these two key notions." And he goes on to say that the ideas in question have been "vaguely founded and modeled upon sensorial evidence" which in bygone days could be regarded as "perennial and safe from attack," but has now been discredited through the momentous discoveries of physics.

But what exactly does Teilhard understand by the terms "immobility" and "substance"? If he speaks of immobility and substance with reference to pure Being, then it is by no means true that "these two key notions" have been "vaguely founded and modeled upon sensorial evidence." We need but to think of Parmenides, for instance, the apostle of immobility and substance one could say, who reputedly went so far as to deny the reality of motion and change on the grounds that these "sensorial" ideas are incompatible with his ontological conceptions. But be that as it may, what the new physics has in fact discredited is not the presumed immobility or substantiality of being as such, but rather the Newtonian idea of atomic particles: little bits of ponderable matter, namely, which supposedly preserve their self-identity or sameness within an ever-changing universe. It is in essence the old atomistic doctrine of Democritus and Leucippus that has been thus disqualified. And it is interesting to note that neither Parmenides, nor Heraclitus, Plato, or Aristotle, nor indeed a single Doctor of the Church has ever upheld that view! On the contrary: what is sometimes termed the perennial metaphysics has always been adamantly opposed to atomism in any of its forms. It was in fact Galileo and Descartes who reintroduced this heterodox ontology, and its subsequent overthrow at the hands of modern physics is thus to be viewed as a partial return to the traditional teaching: a step in the right direction if

---

47. Ibid., p. 182.

you will. Nothing could be more misleading, therefore, than Teilhard's claim to the effect that the findings of physics have disqualified the elements of Christian ontology.

In the final count the doctrine of "participated being" emerges unshaken. Reason alone, perhaps, cannot tell whether that tenet is true; and yet, in the wake of Teilhard's repeated onslaughts one is more inclined than ever to conclude that the teaching may indeed be "perennial and safe from attack."

5
THE OMEGA HYPOTHESIS

IT WAS TEILHARD'S cherished conviction that cosmic evolution must tend towards a universal center of convergence, and that this could be verified on purely scientific grounds. He seems in fact to perceive the discovery of what he termed "Point Omega" as the ultimate recognition of a unified science, a science which has itself converged to that "ultra-physics" of which he sometimes speaks. Yet this epoch-making discovery was for Teilhard but a step towards an even more momentous recognition: the realization, namely, that what appears to the eye of science as a universal center of attraction and confluence—the so-called Point Omega—is in reality none other than the cosmic Christ as conceived by St. Paul! This, then, is the stupendous finding to which Teilhard lays claim.

It hardly needs pointing out that this twofold discovery—or conjecture, more precisely—is central to Teilhard's thought; what is not clear, on the other hand, is what exactly Teilhard has in mind when he speaks of a "convergent" universe. Now the primary sense of the word is of course spatial: "to converge" means to come together at a point. Are we to understand, then, that the cosmos, in its entirety, will eventually collapse into a single point? This is one of those questions which Teilhard does not answer with a simple "yes" or "no." The fact is that he speaks of convergence in multiple ways entailing different senses of the term, which in his mind seem however to be somehow fused; and whereas generally he speaks in metaphor, he does on occasion also employ the term in its primary sense.

It is clear that the gravitational conceptions of Newtonian physics played an important role in Teilhard's thought, as he himself admits when he alludes to "the curiously seductive power that the phenomenon of gravity exerted on my mind while I was still very young,"[1] and when he tells us that "by its gravitational nature, the Universe, I saw, was falling—falling forwards—in the direction of Spirit as upon its stable form."[2] Yet at some point apparently Teilhard began to shift from a purely gravitational notion of cosmic convergence to something far more sophisticated and considerably less clear: "This was no longer universal 'attraction' gradually drawing around itself the cosmic Mass," he tells us, "but that as yet undiscovered and unnamed power which forces Matter (as it concentrates under pressure) to arrange itself in ever larger molecules, differentiated and organic in structure."[3] By now the idea of convergence seems to have metamorphized into a concept of "complexification": "Beyond and above the *concentration-curve*," we are told, "I began to distinguish the *arrangement-curve*...."[4] But does this mean that the notion of a cosmic "concentration-curve" has now been scrapped? Teilhard does not say that, and in fact continues to allude here and there to strict geometric convergence. He does so, for instance, when he delivers himself at considerable length on the so-called "cone of time": as we shall have occasion to see, the geometry of that imagined cone clearly entails a universal confluence. And as if to settle the matter, Teilhard goes so far as to speak of "a growing awareness of the convergent nature of Space-Time" as "the event that characterizes our epoch."[5] I will point out, in passing, that Teilhard could not have been more mistaken on that score, at least; for it happens that the prevailing astrophysical cosmology affirms just the opposite: not that the universe is converging towards a center, but that it has been *diverging* from an initial singularity for the past so many billion years!

Along with the fundamental notion of cosmic convergence one encounters the related idea—again physical and distinctly spatial—

1. HM, p. 33.
2. Ibid., p. 28.
3. Ibid., p. 33.
4. Ibid.
5. FM, p. 96.

of an "irresistible 'Vortex' which spins into itself, always in the same direction, the whole Stuff of things, from the most simple to the most complex; spinning it into ever more comprehensive and more astronomically complicated nuclei,"[6] as Teilhard tells us in graphic terms. And this "structural torsion," as he calls it, is said to result in "an increase (under the influence of interiorization) of consciousness, or a rise in psychic temperature, in the core of the corpuscles that are successively produced."[7]

One tends, of course, to be overwhelmed by this profusion of professedly scientific insights; yet what all this affirms, essentially, is nothing else than Teilhard's celebrated Law of Complexity/Consciousness. Only this time the dual process of complexification and interiorization has been explicitly conceived as the manifestation of an "as yet undiscovered and unnamed power" which presumably resides in the heart of Matter.

But once again there are difficulties. Quite apart from the fact that the stipulated "power" remains admittedly undiscovered, one does know. in any case, that matter as such exhibits a universal tendency to move precisely in the reverse direction: from the complex, namely, to the simple. There is a well-founded thermodynamic law concerning entropy—one of the most solid in all of modern physics—which affirms that a system of particles under the action of physical forces will tend towards a homogeneous state, a state of equilibrium: from order to disorder, as one can say from a kinetic point of view. There are no scientific grounds, moreover, to suppose that this law is somehow abrogated within a living organism. Now it is true, of course, that living organisms tend to complexify during the ascending curve of their life-cycle, and that they maintain a stupendous degree of order; but they do so by ingesting energy from their environment. That is just why we need to eat and to breathe: it takes energy to maintain order. And in the process of maintaining its own order, the organism inevitably causes a corresponding disorder in the environment. We have reason to believe, therefore, that when it comes to the total system—organism plus environment—the ther-

6. HM, p. 33.
7. Ibid.

modynamic law concerning entropy is by no means violated. And this means that it is indeed the universal tendency of matter as such—not to "complexify" as Teilhard believes—but to do the very opposite: to fall into disorder, namely; and the phenomenon of life does not alter this fact.

This leaves two possibilities. One can say that life is inherently a statistical accident of astronomical improbability, perpetuating itself through metabolic devices; and this amounts evidently to the Darwinist claim. Or else one may conjecture that life is the manifestation of a vital principle: a special kind of energy, if you will, different from energy as conceived by the physicist. And this is what not a few biologists have in fact proposed, and what Teilhard himself seems to be saying when he speaks of an as yet undiscovered power, or of such mysterious things as his so-called radial energy. Only in that case the relationship between life and matter would not be a question of evolution—of matter gradually transforming itself into life—but of a ceaseless struggle, rather, between two alien principles tending in opposite directions. The forces of life would thus be engaged in mortal combat, so to speak, with an obstreperous element upon which they have fastened, and which for a time they dominate, only to be vanquished in the end and forced to withdraw at the moment of death, perhaps to return to their native sphere.

These are the two conceptual possibilities, basically, which the law of entropy leaves open; and each in its own way contradicts Teilhard's assumption that "the fundamental property of the cosmic mass is to concentrate upon itself, within an ever-growing consciousness, as a result of attraction and synthesis."[8] And Teilhard is aware of the difficulty; for he immediately goes on to say that "in spite of the appearance, so impressive as a factor of physics, of secondary phenomena of progressive dispersion (such as entropy), there is only one real evolution, *the evolution of convergence*, because it alone is positive and creative."[9] But this is no argument at all! It is simply to deny the validity of the thermodynamic law regarding entropy just because it is incompatible with his own assumptions

8. CE, p. 87.
9. Ibid.

concerning "real evolution." We are asked, in other words, to give up one of the most basic and well-attested laws of physics on the strength of conjectures for which there is no scientific evidence at all.

What is also strange is that, having postulated the existence of a universal tendency on the part of matter "to concentrate upon itself," Teilhard informs us a few pages later that "the whole cosmic Event may be reduced in its essence to one single vast process of arrangement, whose mechanism (that is, the use of effects of Large Numbers and the play of Chance) is governed by statistical necessity."[10] This is strange, I say: for to speak of "Large Numbers and the play of Chance" is to deny implicitly that the phenomenon in question is due to an innate tendency. If a chimpanzee, for example, were set loose at the keyboard of a typewriter, and if it should happen that the resultant text contains an English sentence or two, one would be justified to speak in such statistical terms; but such is obviously not the case when it comes to human authorship. The human typist, one might say, has a tendency to produce English text. And so the formation of sentences is no longer a statistical phenomenon, and has nothing whatsoever to do with "the play of Chance." The same logic, however, applies to that vast process of arrangement envisioned in Teilhard's theory: here, too, the notion of chance is opposed to the idea of innate tendency.

Teilhard himself, moreover, seems at times to realize that his theory of universal convergence does not sit well with the findings of science. But we are given to understand that it is science—and not his theory—that stands at fault. By its very nature, Teilhard maintains, science as such is constrained to deal only with the lower reaches of the evolutive trajectory: "By following science, we have gone no further *than the extreme lower limits of the real*, where beings are at their most impoverished and tenuous"[11] he tells us. Worse still, science is looking in the wrong direction: "What we have been doing is to advance in the direction in which everything disintegrates and is attenuated."[12] But this argument, too, proves to

10. HM, p. 51.
11. SC, p. 28.
12. Ibid., p. 30.

be of no avail. For the charge is first of all inaccurate, seeing that science, for all its categorical limitations (which Teilhard is always willing, in other contexts, to overlook), is nonetheless oriented towards unity. What else is a physical law, after all, than the affirmation of a certain unity expressed in the form of an equation? But if nonetheless science should prove to be incapable of grasping the higher unities to which Teilhard alludes, how then could he claim scientific support for the recognition of Point Omega, the highest unity of all?

Teilhard's reply to this obvious objection is ingenious in the extreme: science, he maintains, "by the very impotence of its analytical efforts, has taught us that in the direction in which things become complex in unity, there must lie a supreme center of convergence and consistence, in which everything is knit together and holds together."[13] Yet rhetoric aside, the question remains how "the very impotence" of scientific analysis could entail the existence of "a supreme center": how on earth do we know that the cosmos at large is converging towards a Point Omega if the issue cannot be resolved on scientific grounds?

But be that as it may, we need to consider yet a third mode or aspect of cosmic convergence: the *psychic*, namely. We encounter this notion squarely in *The Phenomenon of Man* when Teilhard informs us that "all the rest of this essay will be nothing but the story of the struggle in the universe between the unified *multiple* and the unorganized *multitude*: the application throughout of the great *Law of complexity and consciousness*: a law that itself implies a psychically convergent structure and curvature of the world."[14] Yet in place of an explanation as to what all this means, and how "the great *Law*" entails a psychic convergence and cosmic curvature, we are only told, at this juncture, that "we must not go too fast."

---

13. Ibid., p. 34.
14. PM, p. 61.

Later in the book Teilhard gives us to understand that the phenomenon of psychic convergence begins to take place in the human domain, after the trajectory of evolution has crossed "the critical point of reflection." The idea seems to be that a reflective or human consciousness is centered, or ego-centered as one could say: it has not only the capacity but indeed the tendency to relate everything to itself. And so, metaphorically speaking, it gathers up the universe, or a certain small portion thereof at any rate, and concentrates it upon a single point, a single psychic center. There is, then, a "psychic convergence" in that sense.

But the first and perhaps most obvious difficulty with this notion is that there is not just one universal cosmic convergence of that kind, but a vast number of such psycho-physical happenings: one for each reflective center or conscious human being. To speak of a single psychic convergence, at least on a planetary scale, one needs therefore to postulate the concept of a collective psyche; and that is where Teilhard's so-called noösphere comes into play.

The basic idea is that human socialization gives birth gradually to a super-organism, replete with a psyche of its own, i.e., the noösphere. And this happens in compliance with "the great Law." We are told in *Mon Universe*, for example, that "the unification that is being developed so intensely in our time in the human spirit and the human collectivity is the authentic continuation of the biological process that produced the human brain. That is what creative union means."[15]

This, then, is the picture as Teilhard has drawn it innumerable times. And certainly we cannot agree with Henri de Lubac when he writes that "we should not attach too much importance to what we are told about this 'super-organism,' that it will be made up of all human individuals just as the biological individual is made up of cells. Here again, there is no more in this biological language than an analogy, whose shortcomings were recognized by Père Teilhard himself."[16] Now it is true enough that "the individuals that enter into the composition of such a super-organism are not conceived as

---

15. SC, p. 82.
16. *The Religion of Teilhard de Chardin* (NY: Desclee, 1967), p. 208.

ceasing to be so many reflective, personal centers"[17]: Teilhard himself has made this perfectly clear when he tells us repeatedly that collectivization, in the authentically organismal sense, "differentiates" and "super-personalizes" the human cells, and that the entire process is converging towards a "centered system of centers." But this does not mean that Teilhard has retracted his biological claims: how can he? Only an enlarged or extrapolated biology can conceivably provide a quasi-scientific basis that could allow him to speak of a super-organism, an organic noösphere, and an eventual psychic convergence of planetary proportions. What is needed, in fact, is precisely his Law of Complexity/Consciousness, conceived as a *biological* principle: it is the only thing that could lend an appearance of scientific validity to his views regarding the future of mankind. But be that as it may, it is abundantly clear that, in Teilhard's eyes, the idea of a biologically founded noösphere was not just an analogy, whose shortcomings he recognized; if that were the case, he would obviously be deceiving us when he declares, in reference to his usual biological argument: "We cannot, therefore, fail to see that of all living things we know, none is more really, more intensely, living than the noösphere."[18] In a word, on this crucial issue, as on certain other matters, it would seem that Henri de Lubac is trying rather desperately to tone down what his Jesuit confrère has put in print.

But let us go on. The great event which Teilhard needs to postulate is the formation, within the noösphere, of a unique center, or center of centers: "The existence ahead of us," to put it in his own words, "of some critical and final point of ultra-hominization, corresponding to a complete reflection of the noösphere upon itself."[19] This is what present-day humanity is supposedly straining to bring to birth; this is what all the stress, all the Angst and anguish is about. What is straining to emerge "is no longer the simple isolated reflection of an individual upon himself," Teilhard tells us, "but the conjugate and combined reflection of innumerable elements, adjusting

17. Ibid.
18. AE, p. 288.
19. Ibid., p. 290.

and mutually reinforcing their activities, and so gradually forming one vast mirror—a mirror in which the universe might one day reflect itself and so fall into shape."[20] And this will be that psychic convergence of the universe which "the great Law of Complexity implies."

Psychic convergence, therefore, goes hand in hand with the convergence of complexification, so much so that Teilhard leaves us with the impression that the two are not just complementary aspects of a single process, but actually one and the same thing. Thus, after referring to the "complete reflection of the noösphere upon itself," and in the same breath, as it were, Teilhard goes on to say: "We no longer have in the universe nothing but that heartbreaking entropy, inexorably reducing things (as we are still constantly being told) to their most elementary and most stable forms: but, emerging through and above this rain of ashes, we see a sort of cosmic vortex within which the stuff of the world, by the preferential use of chances, twists and coils upon itself ever more tightly in more complex and more fully centered assemblies."[21]

And the passage closes with the following inimitable statement, which may be left to speak for itself: "A world that is in equilibrium upon instability, because it is in movement: and a world whose dynamic consistence is increasing in exact proportion with the complexity of its arrangements, because it is converging upon itself in as many sidereal points as there ever have been, as there are now, and there ever will be, thinking planets."[22]

In keeping with the claim that his theory is science-based, the Teilhardian noösphere has a distinctive geometry: it is supposedly a spherical envelope or aura stretching around the planet Earth. At an earlier stage in the unfolding of his ideas, Teilhard conceived of the so-called biosphere as a "living membrane stretched like a film over

20. Ibid., p. 288.
21. Ibid., p. 290.
22. Ibid.

the lustrous surface of the star which holds us."²³ Some time later he discovered that "there was something more: around this sentient protoplasmic layer, an ultimate envelope was beginning to become apparent to me, taking on its own individuality and gradually detaching itself like a luminous aura. This envelope was not only conscious, but thinking, and from the time when I first became aware of it, it was always there that I found concentrated in an ever more dazzling and consistent form, the essence or rather the very soul of the Earth."²⁴

One has every reason to believe that Teilhard means this to be more than mere poetry. The fact that the spherical-membrane description occurs often enough in a purportedly scientific context shows clearly that it is proposed, not as an account of some kind of mystical experience, but as a sober scientific tenet, unconventional though it may be. Such is obviously the case when we are told that "under the combined force of the multiplication (in numbers) and expansion (in radius of influence) of human individuals on the surface of the globe, the noösphere has for the last century shown signs of a sudden organic compression upon itself and compenetration."²⁵

It must be remembered, however, that the noösphere is supposed to be a *psychic* principle: it is made up of reflective consciousness, or as Teilhard also says, of thought. But he seems not to recognize that for this very reason it cannot be conceived as a corporeal entity: a thing that occupies space. We have already touched on this metaphysical point in an earlier chapter. Consider the visual perception of a landscape, for instance. If we suppose that the resultant percept is itself extended in space—presumably within the brain—it would possess no more unity than, say, a photograph, which consists after all of so many thousand separate black or colored dots. The metaphor of psychic convergence (of a spatial multiplicity gathered into a single center), on the other hand, demands evidently an authentic unity, which is something else entirely. It is an inalienable character-

---

23. HM, p. 32.
24. Ibid.
25. AE, p. 291.

istics, thus, of reflective consciousness *not* to be extended, *not* to be *in space*.

So too we may ask by what conceivable experiment or operational procedure one could localize a given element of consciousness in space: how does one measure the coordinates of a mental image, or determine its height and its breadth? Only a little reflection is needed to see that there actually *are no* such procedures: there cannot be. It is obvious, on the other hand, that a thing which does exist in space *can* be spatially localized; that is to say, it has coordinates which *can in principle* be ascertained, at least to within certain limits of accuracy. It follows that consciousness—or what amounts to the same, its content—is not a thing of that kind. To speak of consciousness or thought as though it were a spatial entity, and do so in purportedly scientific (and hence non-allegorical) terms, is therefore absurd.

Once more, then, let us ask: when Teilhard speaks of the noöspheric envelope as a spherical membrane, which "gradually detaches itself like a luminous aura" from a sentient protoplasmic layer, is he speaking merely in metaphor? It would make little sense to argue that such is the case, seeing that Teilhard himself provides compelling evidence to the contrary. Consider, for instance, the following pivotal statement in *The Phenomenon of Man* which in fact pertains to Teilhard's argument for the existence of Point Omega: "Because it contains and engenders consciousness, space-time is necessarily *of a convergent nature*."[26] We are explicitly told, here, that space-time *contains* consciousness. And not only does Teilhard affirm this misconception, but he uses it as a premise. His argument—which as usual is more implied than expressed—appears to be as follows: *because* consciousness exists in and is engendered by space-time, and *because* in its reflective state consciousness concentrates a certain spatio-temporal content upon a center within itself, *therefore* space-time concentrates itself upon a center. Thus a certain "in-folding" of space-time upon itself takes place. When, therefore, the anticipated "complete reflection of the noösphere upon itself"—which is evidently tantamount to the

---

26. PM, 259.

formation of a unique center of centers—has taken place, that infolding will constitute a convergence of space-time to that supreme center. Such, then, is the implicit argument which permits Teilhard to conclude, with reference to space-time, that "accordingly its enormous layers, followed in the right direction, must somewhere ahead become involuted to a point which we might call Omega, which fuses and consumes them integrally in itself."[27]

What has happened, apparently, is that in the end Teilhard fell victim to his own metaphors: so telling, so life-like, were these metaphors, that eventually he mistook them for reality. For those of us, on the other hand, who can withstand these figures of speech, or better still, dispense with them altogether, there is no further need to concern ourselves with the so-called noösphere and its presumed self-reflection: to see what is going on it will suffice to consider the cognitive act by which we come to know the familiar things in space and time. Now, it is quite clear that nothing actually happens to space, or to space-time, by virtue of this cognitive act: space-time does not develop some mysterious curvature simply because I contemplate a tree or a mountain. Nor does some bit of space compress itself and slip into someone's psyche. There are actually no "radii" of space or space-time converging to a point. There is the cognitive act—which in fact is something science as such is categorically unable to explain[28]—but there is no such thing as a convergence of space-time to a psychic center. The celebrated "convergence to Point Omega"—which has sparked so much euphoria around the globe!—proves finally to be no more than an illusion Teilhard has deftly conjured up by the spell of his justly famous metaphors: he is clearly a master of this art.

But let us get back to Teilhard's theory. It appears that the Teilhardian convergence applies by its very nature to space-time in its total-

---

27. Ibid.
28. Whosoever comprehends what actually stands at issue in even the simplest act of perception has discovered a profound metaphysical truth. On this question I refer again to "The Enigma of Visual Perception," *Science and Myth*, op. cit.

ity: "Caught within its curve," Teilhard explains, "the layers of Matter (considered as separate elements *no less than as a whole*) tighten and converge in Thought, by synthesis. Therefore it is as a cone, in the form of a cone that it can best be depicted."[29] It is not simply a question, thus, of isolated regions detaching themselves from the rest of the cosmos, but of a universal convergence to a single Apex. And this implies that matter "as a whole" becomes eventually spiritualized. Teilhard apprises us of this stipulated fact repeatedly: for example, when he speaks of consciousness becoming "co-extensive with the universe,"[30] and of "a flux, at once physical and psychic, which made the Totality of the Stuff of Things fold in on itself, by giving it complexity, carrying this to the point where that Stuff is made to co-reflect itself."[31]

Yet one knows full well, on seemingly incontrovertible scientific grounds, that our planet will eventually become uninhabitable, and that terrestrial life will cease. All these organic complexities, said to have evolved out of primordial matter in the course of so many millions of years, will eventually be broken down, leaving the Earth every bit as barren and inanimate as it was at the start. And if there be other planets in the universe on which life has evolved, one can say with scientific certainty that the same fate awaits them all.

Now to be sure, this well-known fact poses a formidable problem for the Teilhardian theory, of which Teilhard was of course well aware: "We cannot resolve this contradiction between congenital mortality of the planets and the demand for irreversibility developed by planetized life on their surface by covering it up or deferring it," he tells us quite rightly; "we have finally to banish the spectre of Death from our horizon."[32] But just how *does* one "banish the spectre of Death"? To this all-important question Teilhard responds with another: "Is it not conceivable," he says by way of reply, "that Mankind, at the end of its totalization, its folding-in upon itself, may reach a critical level of maturity where, leaving

29. FM, pp. 91–92 (my italics).
30. PM, p. 309.
31. HM, p. 82.
32. FM, pp. 126–127.

Earth and stars to lapse slowly back into the dwindling mass of primordial energy, it will detach itself from this planet and join the one true, irreversible essence of things, the Omega Point?"[33]

But actually this proposed scenario raises more difficulties than it resolves. One might wonder, for example, what happens to men and women who die *before* mankind has attained the requisite "level of maturity," a consummation which, quite obviously, is not about to happen any time soon. And if it should be the case that the souls of the "prematurely deceased" are able nonetheless to survive, as indeed most of us hope, and as Christianity teaches, what further need is there in that case for the conjectured "critical point"? What has happened moreover, under the proposed auspices, to the so-called Law of Complexity, according to which the psyche is so closely tied to cerebral complexities that the two are virtually inseparable? If consciousness is really "the specific effect" of complexity, how then can the specific effect exist without its cause?

It is hard to avoid the impression that in trying to escape from the difficulty posed by "the mortality of planets," Teilhard has been forced to back away from his erstwhile evolutionist monism to a *de facto* dualism: somewhere along the "evolutive trajectory" he has to admit a certain separation between body and soul, or matter and spirit, contradicting the assumption that the two are simply different faces or aspects of one and the same reality. When all is said and done, a "survival of death" can only be conceived as a parting of the way: there is something that is left behind, and something that moves on. And it is not without interest to note that in a letter written from the Front in 1917, Teilhard himself goes so far as to suggest that "in a way, the whole tangible universe itself is a vast residue, a skeleton of countless lives that have germinated in it and left it, leaving behind them only a trifling, infinitesimal, part of their riches."[34] The idea recurs, moreover, in some of Teilhard's latest compositions; it is clearly in evidence, for example, when he alludes to "this rain of ashes," above which the cosmic vortex "twists and coils upon itself ever more tightly in more complex and more fully centered assem-

33. Ibid.
34. HM, p. 194.

blies," a passage which was written only four years before his death.

Where, then, does Teilhard stand? Does he in fact *take* a definitive stand on this issue? All one can say is that, here and there, he gives us to understand—in a rather indirect manner, to be sure—that some portion of terrestrial matter gets caught up in the vortex of complexification, and having crossed a certain threshold, becomes permanently transformed into thought, while the remainder is swept away by the downward current of entropy.

But how does this square with the notion of a single Apex, and with the idea that "the Totality of the Stuff of Things" folds in upon itself? What will happen to the less fortunate portion of terrestrial matter which ends up in the current of "that heartbreaking entropy"? Will it, too, become eventually "hominized"? Nowhere does Teilhard actually say that it does. And yet, when he speaks of a spiritualized mankind "leaving Earth and stars to lapse slowly back into the dwindling mass of primordial energy," he does seem to suggest, however faintly, that the "mass of primordial energy"—by which he presumably means the amount of physical energy in the universe—is gradually diminishing, and may eventually be reduced to zero. Are we then to suppose, perhaps, that the lapsing Earth and stars will all be somehow recycled, so that every particle in the universe will eventually be complexified and hominized? Is there to be eventually a single ascending vortex without any compensating "rain of ashes"? And what about the notion, implicit in all of this, that the formation of a human soul, or its departure from the body, is associated with a certain mass defect: if such be the case, why are we not told of this momentous fact in unequivocal terms? And why has no such effect ever been observed?

Let us not belabor the point. One has every reason to believe that the total amount of matter (or better said, physical energy) within the cosmos is strictly conserved; the most exact of our sciences guarantees this. We are left, therefore, with only two conceivable alternatives: one can say, with the materialists, that there is no such thing as soul and immortality; or else one can admit the traditional dualism of body and soul. But it seems that Teilhard wants to have it both ways: he wants soul and immortality, but on an essentially materialistic basis. Small wonder that he finds the going tough!

Enough has now been said regarding the Teilhardian notion of cosmo-convergence to show that the theory is not by any means well founded, to say the least. One is astonished that Teilhard could have promulgated such nebulous and incoherent speculations dogmatically as a scientific truth, when in fact there is actually not a shred of evidence in support of his hypothesis (assuming that he has one): "the evidence that Science provides"[35] exists nowhere except possibly in Teilhard's incredibly fertile imagination. As George Gaylord Simpson, the well-known evolutionist and friend of Teilhard de Chardin, points out in his review of *The Phenomenon of Man*, the Omega doctrine has no basis in scientific fact. "One cannot object to the piety and mysticism of his book," he goes on to say, "but one can object to its initial claim to be a scientific treatise, and to the arrangement that puts its real premises briefly, in part obscurely, as a sort of appendage after the conclusions drawn from them."[36] The fact is that Teilhard has promulgated his private mystical theories under the colors of Science.

The question remains how Teilhard's doctrine stands from a *theological* point of view: must it really be admitted that "one cannot object to the piety and mysticism of his book"? And on this score, too, there is much to be said: it is in fact a question which will concern us for the remainder of this monograph. The problem has many aspects that shall need to be dealt with successively, starting with what is no doubt the crucial issue: Teilhard's identification of Point Omega with Christ.

To begin with, it could certainly be argued that inasmuch as the so-called "Omega Point of science" has turned out to be fictitious, it actually makes no sense to ask whether that nonexistent Point can be identified with Christ. And yet it will not be without interest to observe how Teilhard conceives of the Incarnate Lord.

"In a Universe of 'Conical' structure," we are told, "Christ has a place (the apex!) ready for Him to fill, where His Spirit can radiate

---

35. HM, p. 91.
36. *Scientific American*, April 1960, p. 206.

through all the centuries and all beings."[37] Now, to begin with, the "apex" of that so-called "cone of Time" can be neither a *place* (i.e., a spatial locus), nor an *event* (i.e., a point of space-time), inasmuch as it must evidently be situated "outside" the space-time continuum, or on its "boundary" as one may say. Nothing whatsoever, therefore, can be "located" at that apex in a *physical* sense. Nor is it conceivable that a physical entity could somehow be moved into that apex: in fact, the only possible way to get there from within the space-time continuum would be to wait till "the end of time," when all the so-called world-lines will (supposedly) merge in that singularity.

It needs to be understood, moreover, that Teilhard is speaking indeed of the Incarnate and Risen Christ, as distinguished from the Logos as such: the eternal and "pre-cosmic" Word of God. It is in fact doubtful that Teilhard *ever* conceives of Christ in His pre-cosmic identity as the Second Person of the Holy Trinity. To put it in the language of the Gospels, he seems to be concerned *exclusively* with the Son of Man as distinguished from the Son of God. But be that as it may, his present meaning is perfectly clear: Teilhard is telling us that the Incarnate Christ, by virtue of His Resurrection, has positioned Himself at the Apex of the postulated Time Cone. But this actually makes no sense at all! For as we have just seen, it is certainly not by entering into space-time that the imagined Omega Point is to be reached.

Nor can it be said, by way of rejoinder, that Teilhard conceives of these matters in some high and exclusively mystical sense immune to scientific analysis; for it is indeed the very point of his theory, after all, to amalgamate theological and scientific conceptions. We are perfectly within our right, therefore, to take him at his scientific word.

As a matter of fact, Teilhard goes out of his way to stress the physical and supposedly scientific nature of his Christological speculations. "Starting from the evolutive Omega," he tells us, "at which we assume Christ to stand, not only does it become possible to conceive Christ as radiating *physically* over the terrifying totality of things, but what is more, that radiation must inevitably work up to a maximum of penetrative and activating power."[38] But here again Teilhard

---

37. FM, p. 98.

misses the mark: for it is a basic and quite elementary principle of physics that *velocity vectors point into the future*. And this implies that the postulated Apex (if it exists) would be the one Point from which *no radiation whatsoever* could penetrate into the universe, let alone "work up to a maximum of penetrative and activating power." Christ would thus have positioned himself at the worst possible spot.

Nor does it make any more sense when Teilhard speaks of the Ascended Christ as having been "raised to the position of Prime Mover of the evolutive movement of complexity-consciousness,"[39] For nothing could be more obvious than that a Prime Mover, if he exists at all, must be there from the start. A Prime Mover who takes up his post, so to speak, in the middle of the evolutive process is an absurdity. Or are we perhaps to suppose that there was at first another Prime Mover, who was later replaced by the Risen Christ?

It is certainly correct theologically to speak of the cosmic Christ as a universal Center of attraction and influence, or of "radiation" if you will. But Teilhard evinces an astonishing lack of theological comprehension when he maintains that this radiation is *physical.* Though Christ did assume a physical body, His "radiation" is primarily spiritual: it is in fact ultimately the Holy Ghost. This is the "fire" Christ came "to cast upon the earth"[40]; and whosoever refers to that Fire as a "physical radiation" has missed the point.

The fact is that Christianity knows nothing of a "convergent universe" in the Teilhardian sense. It most certainly does not envision a conical space-time, presided over by an Apex at which, supposedly, all world-lines meet at the end of time. Yet it does also envisage a mysterious Point where God and cosmos meet. What then—or better said, *where*—is that Point?

Contrary to a certain misconception, the God of Christianity is not "transcendent" in an exclusivist sense: He is not in reality an

---

38. HM, p. 95 (Teilhard's italics).
39. Ibid.
40. Luke 12:49.

"extrinsicist God" as Teilhard at times complains. Authentic Christian doctrine affirms that God is not only transcendent, but *immanent* as well. "Since God is the universal cause of all being," writes St. Thomas Aquinas, "in whatever region being is to be found, there must be the divine presence."[41] Yet of course not as an object, not as a "thing"!

But then, if God is present everywhere, how can there be a single "Point where God and cosmos meet"? Is it possible, in other words, to conceive of an *omnipresent* Point? Now, as every mathematician will readily understand, it is quite easy to do so *formally*. One need only define a 5-dimensional cone which contains our 4-dimensional space time as its base, and endow that 5-dimensional cone with a non-Euclidean metric such that the distance from its apex P to any point X in space-time is precisely *zero*. Since P is thus "present at" X, we have constructed an omnipresent point.

The Fathers of the Church, of course, knew nothing of such a geometry. Yet they understood the *idea* which the aforesaid model exemplifies, which is in fact crucial to their theology. One may conceive of that omnipresent point P in many ways: for instance as the *nunc stans* or "now that stands" of the Scholastics, which is indeed none other than the "νῦν" of St. Paul when he declares: "Behold, *now* is the accepted time; behold *now* is the day of salvation."[42] One could write a whole treatise on the theology and metaphysics of that Point! And I will note in passing that the conception of such a universal Center is in fact to be found in virtually every metaphysical tradition of mankind.[43]

Let us then consider how that Center—which we take to be the authentic Point Omega—relates to the Teilhardian version. And this is most readily accomplished by way of our mathematical model: the 5-dimensional cone, namely, which differs from Teilhard's Time Cone, first of all by virtue of the fact that it embodies an extra dimension: the dimension of *verticality* one can say. That dimension

---

41. *Summa Contra Gentiles*, III.68.
42. 2 Corinthians 6:2.
43. See especially A.K. Coomaraswamy, *Selected Papers* (Princeton University Press, 1977), vol. 1, pp. 415–544; vol. 2, pp. 220–230.

has evidently disappeared in Teilhard's so-called theology. At the same time a rotation of axes has taken place—as we have noted before—which in effect transforms "the above" into "the ahead." The time-axis of space-time has thus replaced the dimension of verticality: that is the trick which has apparently befuddled and misled millions.

But let us get back to the unfalsified doctrine. I would like to point out that the metaphysical conception, which we have thus exemplified in mathematical terms, is in a way implicit in virtually every article of Christian belief. Abstract and abstruse though that conception may be in its own right, every aspect of the spiritual life hinges in fact upon the truth it enshrines. From the parables of Christ to the Patristic writings or the canons of Christian art: everywhere we encounter that truth. A simple observation, perhaps, may help to make this clear: the "model" which we have described in such forbidding terms is visibly exemplified, for all to see, by the "vault of heaven" and by the dome in sacred architecture, which may be thought of as a replica of that vault. Here, in the cathedral of God or the basilica of man, we can actually behold the ascending radii soaring upwards high above the ground on which we stand, and see them converging to a central point, at which the august figure of the Pantocrator can be imagined or perceived. The faithful, moreover—even the simplest among them—understand full well that the "distance" from this plane of earth to where Christ stands is not to be measured in meters or miles: it is a *spiritual* distance, which the "pure in heart" can cover in a trice.

The cosmos, then, according to Christian conception, does have an Apex. And that is in fact the Point from which "Heaven and all Nature hang," to put it in Dante's magnificent words. "Here every where and every when are focused," the poet goes on to say: for wherever we may be, and whatever the moment of cosmic time, we *do* confront that Point. It does not fluctuate, does not move: the world moves, but that Point remains ever fixed. For it is in truth "the pivot around which the primordial wheel revolves."[44]

---

44. *Paradiso*, 28.41, 29.12, and 13.11, respectively.

Here, then, is the authentic Point Omega, which is moreover unique: no other such Point exists. And that Point turns out to be—not the conjectured *End* of the universe—but its ever-present *Center*. It is thus to be encountered, not as the endpoint of an evolutive trajectory, but precisely in the here and now. We need not, after all, wait so many million or billion years to get there: happily it can be accessed much sooner than that. What is more, the faithful do in fact access that Point daily: for example, when they pray earnestly to their Father *"who art in heaven."*

6

THE GOD OF EVOLUTION

THERE HAS BEEN MUCH debate regarding Teilhard's theological orthodoxy. It is of course necessary, first of all, to ascertain precisely where Teilhard stands on the basic issues; and this, as we have seen, is not an easy task. "Imprecision or contradiction in definition is one of the constant problems in the study of the Teilhard canon"[1] observes George Simpson. It is hardly surprising, therefore, that opinions have varied greatly as to what, exactly, Teilhard did say regarding the nature and attributes of God.

Obviously he does exhibit pantheistic tendencies; but just how far does he go? There has been ample discussion on this issue, but no general agreement. The learned Dominican Guerard de Lauriers, for instance, maintains that Teilhard espoused "a veritable metaphysical monism," a monism "so radical that it removes being,"[2] whereas other theologians, beginning with Henri de Lubac, plead the case of Teilhardian orthodoxy. "Père Teilhard, one need hardly say, believed in God," writes de Lubac in answer to de Lauriers; "but he believed also, and affirmed, that, transcending the world, 'God could dispense with the world,' that he was self-sufficing; that the inevitability that we see in the world is only 'a consequence upon the free will of the Creator.' That in itself is enough to dismiss the accusation."[3]

But in fact it is not: for it happens that elsewhere Teilhard flatly contradicts these orthodox-sounding affirmations which his distinguished confrère has adduced. In an essay entitled "Suggestions for

1. *Scientific American*, April 1960, p. 204.
2. "La demarche du Père Teilhard de Chardin," *Divinitas*, vol. 3 (1959), p. 227.
3. *The Religion of Teilhard de Chardin* (NY: Desclee, 1967), p. 195.

a New Theology," for instance, he makes it a point to reproach the traditional doctrine precisely for its belief in the absolute self-sufficiency of God: "God could, it appeared, dispense with the universe,"[4] he charges. This time-honored belief, he insinuates, has now become outmoded and needs to be given up. So too we read, in one of his latest works: "In truth, it is not the sense of contingence of the created but the sense of the mutual completion of the world and God which gives life to Christianity."[5] In other words, even as the world has need of God, so too God has need of the world: a far cry indeed from the dictum "God could dispense with the world" cited by de Lubac.

Nor does the case for Teilhardian orthodoxy stand any better when it comes to the stipulated "free will of the Creator." Here too de Lubac seems to forget that his confrère has often enough expressed himself on the opposite side of the issue. He does so, for instance, in one of his very early compositions, dated 1919, when he asks: "In making God personal and free, Non-being absolute, the Creation gratuitous, and the Fall accidental, are we not in danger of making the Universe *intolerable* and the value of souls (on which we lay so much emphasis!) inexplicable?"[6] One can hardly fail to sense how much of Teilhard's later theological thought is already implied in this fascinating sentence, written just seven years after his ordination.

Which brings us to another question: the "evolution," namely, of Teilhard's theological beliefs. To be sure, the Teilhardian writings, spread out as they are over a period of some forty years, do exhibit an unfolding or development of ideas. So far as theological questions are concerned, moreover, they manifest an ever more overtly unorthodox stance. And as a matter of fact, Teilhard himself alludes, with evident satisfaction, to a progressive estrangement from the traditional Christian teaching. In *The Heart of Matter*, for instance, written five years before his death, he speaks condescendingly of an earlier period in his life, during which he was still committed to various traditional conceptions, still subject to "those odd

---

4. CE, p. 182.
5. Ibid., p. 227.
6. HM, p. 219.

effects of inhibition that so often prevent us from recognizing what is staring us in the face."[7] It was during this period, moreover, preceding his "emancipation," that Teilhard composed his most nearly orthodox essays: the writings which are unfailingly cited by de Lubac and others as proof of Teilhard's theological innocence. But they are missing the larger picture: missing the point. "I can see quite clearly" Teilhard tells us, looking back upon these early years, "how the inspiration behind 'The Mass on the World' and *Le Milieu Divine* and their writing belong to that somewhat self-centered and self-enclosed period of my inner life."[8]

On the other hand, we must not make too much of this alleged change of outlook. It was not a radical change, not by any means a reversal, but rather a recognition and an embracing of tendencies which had been there from the start. In some of his earliest essays, as we have seen, Teilhard had already voiced distinct misgivings concerning fundamental tenets of orthodox theology, such as the absolute freedom of God and the gratuitous nature of creation. The seeds of divergence, clearly, were present from the start. But they needed time to develop. One must also remember that in his younger years Teilhard was of course less sure of himself, and not as free in the expression of his less than orthodox views. Yet one also has reason to surmise that a certain lingering and fragile orthodoxy—another voice within—may have survived the transformation: one finds occasional passages even in his later works which tempt one to believe that such was the case. Are these still to be counted, perhaps, among those so-called "odd effects of inhibition"? What we do know for certain is that Teilhard did become progressively less orthodox and more radical in the expression of his beliefs: unquestionably to the point of explicit heresy.

Every theologian worthy of the name has conceived of God as immutable: only created things—only creatures—are subject to

---

7. Ibid., p. 52.
8. Ibid.

change. "They shall be changed, but thou art the same."[9] Not that God remains somehow fixed, like a stone; the point, rather, is that He is unaffected by time. His Being, unlike our own, is not spread out, as it were, over a temporal continuum. How could it be? How could the Author of time be subject to change? "Before Abraham was, I am"[10]: not "I was," but "*I am.*" Clearly, by this startling use of the present tense, Christ has proclaimed a capital truth. We are given to understand that the "I am" which He declared to the Jews is none other than the "*ego sum qui sum*" proclaimed unto Moses when he asked to know God's Name.[11] What stands at issue is a mode of Being beyond the pale of time, a mode that belongs to God alone. And let us by all means observe that so far from being simply a matter of "academic" interest, that is a truth vital to the Christian Faith. There is in fact a connection, as St. Augustine points out,[12] between the pronouncement of John 8:58 and the dire warning sounded in John 8:24: "If ye believe not that I am [*ego sum*], ye shall die in your sins."[13]

This having been noted, it will certainly be of interest to observe what an "evolutionist theology" has to say on that issue. Where exactly does Teilhard stand? Does he submit to the orthodox view: namely that God is "above time," and is thus indeed immutable? It would be absurd to claim that he does embrace that view: not, certainly, after he had freed himself from "those odd effects of inhibition" which, in his earlier years, had supposedly prevented him from seeing the light. As Teilhard himself observes: "I failed to understand that as God 'metamorphized' the World from the depths of matter to the peaks of Spirit, so in addition the World must inevitably and to the same degree 'endomorphize' God." To which, by way of clarification, he adds: "As a direct consequence of

---

9. Psalm 101:27–28.
10. John 8:58.
11. Exodus 3:14.
12. In *Joannis Evangelium*, 38:10.
13. It seems strange that the *ego sum* (or *ego eimi* in the Greek text) should have been rendered by the phrase "I am he," as in fact it has been in so many English translations. Not only is this translation obviously incorrect, but it obscures the metaphysical sense of this crucial declaration.

the unitive process by which God is revealed to us, he in some way 'transforms himself' as he incorporates us."[14]

Here, in his treatise entitled *The Heart of Matter* which appeared five years before his death, Teilhard seems indeed to have cast off his former "inhibitions." He goes out of his way, in fact, to make his radically unorthodox point with the utmost clarity: "All around us, and within our own selves, God is in process of 'changing'" he declares; "his brilliance increases, and the glow of his coloring grows richer."[15] This is no doubt that "mutual completion of the world and God which gives life to Christianity" alluded to in one of his last essays. Not just the world, but God, too, is changing, and indeed "for the better": that is clearly the message. "I see in the World a mysterious product of completion and fulfillment for the Absolute Being himself"[16] we read. Never mind that Christ Himself declares otherwise: "Be ye perfect, even as your Father in heaven *is* perfect": present tense, once again. But perhaps this teaching pertains to an earlier phase of human evolution, antedating Darwin and "the discovery of Time." In any case, Teilhard informs us in no uncertain terms that in fact "the Absolute Being" is not yet fully perfect, and that God Himself depends upon Evolution for His "completion and fulfillment"—as if this fact alone would not suffice to render "the Absolute Being" less than absolute!

It must not be supposed, moreover, that when Teilhard attributes change or transformation to God, he is referring specifically to the human nature of Christ, or perhaps to His Mystical Body. If that were the case, why would he speak of God and "the Absolute Being" instead of Christ? Granting that Teilhard is not always precise in his use of theological terms, it must nonetheless be assumed that he knows very well how to distinguish between the Absolute and the Incarnation. What settles the matter, however, is that the conception of a mutable or "evolving" God is in fact demanded by Teilhard's fundamental outlook: his rejection, namely, of the traditional Christian doctrine concerning creation and "participated being,"

---

14. HM, pp. 52–53.
15. Ibid., p. 53.
16. Ibid., p. 54.

together with his theory of "creative union" which is supposed to replace these "antiquated" ideas. That "there is ultimately no unity without unification"[17] can only mean that God Himself has no unity apart from what He derives by way of the evolutive process. And this entails that God's unity is not yet complete: He too must await "the end of the world" when all shall be fulfilled.

Most assuredly these apodictic claims repudiate the position Teilhard had expressed early on in "The Mass on the World": they flatly contradict the notion that "the world travails, not to bring forth from within itself some supreme reality, but to find its consummation through a union with a pre-existent Being."[18] Or are we perhaps to suppose that even here, in this seemingly orthodox affirmation, there lurks a hidden implication: that perhaps the pre-existent Being becomes somehow enlarged by the union that is to come? This too is conceivable; for it will be recalled that in 1919, four years before "The Mass on the World," Teilhard had already voiced opposition to the idea of "participated being." Yet the fact remains that only much later, according to his own testimony, did he come to realize "the full truth": that just as God "metamorphized" the world, "so in addition the World must inevitably and to the same degree 'endomorphize' God." Herein lies the quintessence of his doctrine, the culmination of his thought. And by his own admission he was for long prevented from recognizing this reciprocity between God and world on account of a lingering adhesion to the traditional outlook, from which he had not yet emancipated himself.

This does not mean, of course, that Teilhard was ever fully orthodox. From the start it seems to have been his tendency to conceive of God in a distinctly unorthodox sense as "the Evolver." In an essay entitled "The Modes of Divine Action in the Universe," for example, written in 1920, he maintains that

> God's power has not so free a field for its action as we assume: on the contrary, in virtue of the very constitution of the participated being it labors to produce . . . it is always obliged, in the

17. SC, p. 184.
18. HM, p. 129.

course of its creative effort, to pass through a whole series of intermediaries and to overcome a whole succession of inevitable risks—whatever may be said by the theologians, who are always ready to introduce the operation of the *'potentia absoluta divina'*.[19]

Teilhard seems to assume that the conception of God's omnipotence was the invention of overzealous theologians, forgetting that the idea is thoroughly Biblical: "He spake, and it was done; he commanded, and it stood fast" declares the Psalmist.[20] Nothing, in fact, could be more incongruous for a believing Christian than the notion of a God who "labors to produce," and "is always obliged" to take chances. One sees that even at this early stage in the unfolding of his theological speculations, Teilhard's deviation was already well under way. The Absolute had already become relativized and diminished in relation to the universe. From the outset Teilhard seems bent upon blurring the demarcation between God and world, to the point where it is no longer clear whether God is indeed the Creator of the universe. In fact, we are told that He is not: "Properly speaking, God *does not make:* He *makes things make themselves.*"[21] True to his evolutionist convictions, Teilhard conceives the creative act as a temporal process in which the creature has its share. It is the Darwinist conception, with the proviso that "behind the scenes" there is a God exerting an influence. But not enough to interfere substantially with the Darwinist workings of the evolutive process: for Teilhard is careful to point out that "God never acts except evolutively."[22] Whatever it may be that God contributes to the evolutive process, the random gropings and the play of blind chance remain: "theistic" or not, what confronts us is still the Darwinist scenario. And this is for Teilhard a matter of paramount importance: a basic principle, in fact, which "seems to me to be necessary, and all that is necessary, to modernize and give a fresh start to Christianity."[23]

19. CE, p. 31.
20. Psalm 32:9.
21. CE, p. 28.
22. Ibid., p. 16ff.
23. Ibid.

The culmination, however, was yet to come. Teilhard had long believed that God "metamorphizes" the world; but only in the closing years of his life did he come to realize the world, too, acts upon God: that in fact it must "inevitably and to the same degree 'endomorphize' God." The Teilhardian God ceased at this point to be simply "the Evolver," and became, at least in part, a product or result of the evolutive process.

"I see in the World a mysterious product of completion and fulfillment for the Absolute Being himself": this statement was deemed heretical by the Roman censors commissioned to examine the Teilhardian writings in 1948. But whereas this verdict seems to be more than warranted, the point has nonetheless been disputed by theologians of a liberal bent. Teilhard's erudite editor, for instance, has taken it upon himself to argue in a lengthy footnote that the censured Jesuit was right all along: the notion that God Himself is somehow "completed" by a transfigured world, says he, is in fact orthodox. Even Cardinal Pierre de Bérulle, we are told, has implied as much. But what exactly did this venerable and saintly theologian say? Here is the passage in question:

> God the Father, who is the fontal source of the Godhead... produces two divine Persons in himself. And the Son, who is the second producing Person in the Godhead, concludes his productiveness in a single divine Person. And this third Person, who does not produce anything eternal and uncreated, produces the incarnate Word. And this incarnate Word... produces the order of grace and of glory which ends... in making us Gods by participation.... This completes God's communication in himself and outside himself.[24]

This is the excerpt from *Les Grandeurs de Jesus* (1623) which supposedly legitimizes Teilhard's contention that God "'transforms himself' as he incorporates us." But it happens that this passage

---

24. Quoted in HM, p. 79n.

(which is as beautiful as it is orthodox) actually shows the very opposite. For in affirming that the Holy Spirit "does not produce anything eternal and uncreated," de Bérulle implies that the Triune God is Himself "eternal and uncreated." This metaphysical fact, however, entails that the aforesaid God is not subject to change, transformation or evolutive increase: for inasmuch as He is eternal, God is exempt from the condition of time, and being uncreated, He is *ipso facto* exempt from any conceivable effects of "creative union." The "production," therefore, to which de Bérulle alludes, beginning with the Incarnation, refers to "the order of grace and of glory." Here, and here alone, time enters the picture. There does exist an evolution, if you will: something *is* being transformed and perfected. Yet what is undergoing these transformations—what suffers change—is not God, but the creation: it is *we* who must grow, it is *we* who must be made perfect, "even as your Father in heaven *is* perfect." And that, too, is the reason why de Bérulle distinguishes between the communication of God "in Himself and outside of Himself." The order of grace and glory, for all its splendor, is yet to be counted as something "outside" of God. Christianity insists on this point: even in the beatified state the demarcation between God and creature is not obliterated. There is a wondrous union, but not an identity: and it is this razor-sharp distinction that differentiates Christian orthodoxy from pantheism in any of its forms. Now this is all very basic; and both Teilhard and his zealous editor should have understood and acknowledged what the Church has always affirmed.

The fact is that *theologically* Teilhard has not a leg to stand on. And for this reason alone it was necessary to present his wide-ranging speculations in a scientific garb. The credentials of Science were in fact needed on two counts: first to disqualify the old theology, and then to validate the new. We are obliged supposedly to choose between a pre-scientific doctrine which is no longer tenable, and a new theological outlook consonant with the latest scientific discoveries. Like Freud, Jung, and other intellectual superstars of our time, Teilhard makes it a point to present himself first and foremost as an empiricist, a man of science. He clearly does not wish to be perceived as a theologian in the established sense. The image he projects is rather that of a pioneer: the first "scientist-theologian" if you will. He presents himself, in other words, as the inaugurator of a brand new theology: a scientific theology, no less, which is no longer bound by the old rules nor subject to censure on the strength of traditional norms. "Before all creation, proclaims the Scholastic, the Absolute existed in its fullness" he declares by way of highlighting precisely what he denies. The point being that his own doctrine is different, not just on this issue, but radically, that is to say, in its very basis and method: "For us who are simply trying to construct a sort of ultra-physics, by combining the sum of our experiments in the most harmonious way, the answer to the problem is not so positive. From the empirical point of view there is no pure act but only a final term to which the serial bundle that envelops us is converging."[25]

Yet ultra-physics notwithstanding, the fact remains that Teilhard does ceaselessly theologize—for the simple reason that he speaks of God. And in so doing, he clearly departs from "the empirical point of view" to which he pretends to confine himself; for nothing could be more obvious than the fact that the idea of God is *not* an empirical notion. It is from the start out of the question, therefore, that Teilhard's quasi-theological speculations could conceivably be validated on empirical grounds.

It is to be noted that Teilhard is right, of course, when he claims that "from an empirical point of view there is no pure act"; yet it would have been more enlightening to say that the question makes no empirical sense in the first place. But not only is there no pure act: it happens that "from an empirical point of view" there is also no such thing as "a final term to which the serial bundle that envelops us is converging." For as we have already had occasion to see, there is actually no empirical evidence whatever in support of Teilhard's Omega hypothesis. From the start the celebrated Point Omega has been nothing more than a quasi-theological notion masquerading in scientific garb. Teilhard was less than candid with his readers in that regard, as even his friend George Simpson has pointed out. He misleads us when he speaks of "combining the sum of our experiments"—as if by some scientific calculation, too

---

25. HE, p. 70.

formidable for laymen to grasp, an expert in "ultra-physics" could infer the existence and properties of that stipulated Point! Is it a case of conscious deception, then? Or may we suppose that these baseless claims have been put forward in good faith by an individual incapable of distinguishing between scientific fact and poetic flights of the imagination?[26]

This is ever the unanswerable question. What is clear, on the other hand, is that the postulated Point had been earmarked from the start to serve as the god-term of the Teilhardian doctrine. From the outset Teilhard had invested that imagined entity with a plethora of quasi-divine attributes, thinly disguised in scientific-sounding terminology. Knowingly or unknowingly, as the case may be, he fabricated a scientistic fantasy which came more and more to bear the imagined features of "the cosmic Christ." The great task, now, was to expand that notion into a full-fledged "science-based" theology.

But of course this could not be accomplished without negating orthodox theology at virtually every turn. It was therefore necessary first of all to undermine the authority of the theological tradition; and here, too, as we have noted before, the credentials of Science come into play. And again the fact remains that there *are* no empirical grounds on which to dispute theological propositions. Here too there can be no *bona fide* argument, which is to say that once again Teilhard is forced to "cheat with words" to use Medawar's expression. We are told, for example, that so long as men believed in "a static world" it was possible to think of the Creator as "structurally independent of his work," whereas today, knowing that we find ourselves in an evolutive universe, "God is not conceivable (either structurally or dynamically) except insofar as he coincides with (as a sort of 'formal' cause), but without being lost in, the center of convergence of cosmogenesis."[27] One is hard pressed, of course, to comprehend how exactly God could "coincide" with Omega "without being lost" therein: one would think that if two terms coincide there is an end of the matter. But be that as it may, we are told in

---

26. This would be essentially tantamount to Medawar's claim that "before deceiving others he has taken great pains to deceive himself" (*Mind*, vol. 70, p. 99).

27. CE, p. 239.

effect that the old theology is somehow tied to a static conception of the world which has now been disproved, and that it needs consequently to be replaced by a theology which squares with the fact that we find ourselves in an *evolutive* universe. This is the suggestion, the innuendo. But let us recall, in the first place, that the idea of universal flux was evidently familiar, not only to Heraclitus, but to many of the Greek and Latin Fathers, who nonetheless staunchly believed in a "structurally independent" God. There is no logical conflict whatever between the idea of a transcendent, eternal, and immutable God and a world in perpetual flux or in evolutive progression. No matter how irrefutable and scientific what Teilhard terms "the truth of Evolution" may be, his claim to the effect that this discovery invalidates the tenets of orthodox theology is patently false.

As Teilhard himself admits, his doctrine is indeed pantheistic.[28] The die was cast when he rejected the Judeo-Christian idea of creation, or what amounts to the same, the concept of participated being. And Teilhard knows it: "Since God cannot be conceived except as monopolizing in himself the totality of being," he tells us, "then either the world is no more than an appearance—or else it is in itself a part, an aspect, or a phase of God."[29] The alternative is logically indisputable, despite the fact that "appearance" could of course mean many things: the Hindu *māya*, for instance, as popularly conceived. From a Christian perspective, on the other hand, the world is not simply an "illusion," or a kind of "cosmic dream"; it is rather "a lower or secondary form of being gratuitously drawn from 'non-being' by a special act of transcendent causality" as Teilhard rightly points out. And this "lower or secondary form" is in fact tantamount to "participated being," a notion which he castigates and categorically rejects, as we have noted in Chapter 4. Now the logical

28. See, for instance, CE, p. 171, where Teilhard extols what he terms "Christian pantheism."
29. SC, p. 180.

consequence of this rejection is clear: by virtue of the aforementioned alternative it implies that the world has become "a part, an aspect, or a phase of God," which of course is pantheism. It means that God has become the totality, all that exists.

Admittedly this is a rather popular and enticing conception nowadays: Teilhard may be right when he speaks of a mounting "passion for the Whole."[30] What is more, the idea has of late received considerable support from scientific quarters. One must remember that after more than two centuries of unmitigated Newtonian materialism science has at last discovered that "the whole is more than the sum of its parts," which is precisely what distinguishes quantum mechanics from Newtonian physics. An entirely new *Weltanschauung* has thus emerged. As David Bohm has put it:

> One is led to a new notion of unbroken wholeness which denies the classical idea of analyzability of the world into separately and independently existing parts.... Rather, we say that inseparable quantum interconnectedness of the whole universe is the fundamental reality, and that relatively independently behaving parts are merely particular and contingent forms within this whole.[31]

And from here it is but a small step to a holistic pantheism. Having come to the realization that "the whole" is indeed more than the sum of the parts, what could be more natural for individuals of a scientistic bent than to conclude that the biggest conceivable "Whole" could be none other than God?

But the fact remains that the stipulated identification is incompatible with Christian theology. Nor, for that matter, is it concordant with the traditional tenets of the East—notwithstanding all that Fritjof Capra has written on that score.[32] For whatever differences

---

30. CE, p. 65.
31. D. Bohm and B.J. Hiley, "On the intuitive understanding of nonlocality as implied by quantum theory," *Foundations of Physics*, vol. 5 (1975), pp. 96 & 102.
32. Capra claims that the new physical *Weltanschauung* agrees substantially with the mystical teachings of China and India. He has argued the case eloquently in *The Tao of Physics* (NY: Bantam, 1977), a book which despite its serious and at times demanding content has been something of a "best-seller" for years.

there are between the Christian and Oriental doctrines, both most certainly maintain that God is infinitely more than the universe, no matter how holistically the latter may be conceived.

There is a sense, of course, in which one can legitimately speak of Deity as "the Whole"; God is indeed "the Whole" insofar as He is conceived "as monopolizing in himself the totality of being," to put it in Teilhard's phrase. But He becomes the "*cosmic* whole" only if one has rejected, or failed to grasp, the crucial distinction between absolute and contingent being. And this rejection is decidedly unorthodox: for in one way or another that distinction has been sharply drawn within every major traditional school, be it in the East or in the West.

There is then a categorical distinction between the divine and the cosmic "whole," even as there is a corresponding distinction between absolute and contingent *being*. This can perhaps be most readily understood if we remember that God is the Architect or Lawgiver of the universe; as we read in the Book of Proverbs, He has "set a compass upon the face of the deep."[33] Cosmic or contingent being, then, is "the measured," whereas absolute being is "unmeasured" and indeed unmeasurable, which is to say that it is *absolute*. And let us note that this is precisely the reason why the former pertains to the domain of science whereas the latter does not: only "the measured" can be known by scientific means. Science, by its very nature, is restricted to the cosmic domain. This is all it knows and all it *can* know: the *unmeasured* and *unmeasurable* will forever elude its grasp. And this holds true not only for science, but indeed for human knowing as such. After all, we know by way of *concepts*; but what is a concept if not a bound or limitation of some kind? The *being* of God proves thus to be humanly unknowable. And this, too, has always been recognized by the wise. "What, then, can I do?" exclaimed St. Augustine. "What that existence is, let Him tell, let Him declare it within; let the inner man hear, the mind apprehend

---

33. Proverbs 8:27. It is not without interest to note that a similar text is to be found in the Rig Veda (VIII.25.18), which reads: "With His ray He has measured heaven and earth." It appears that this doctrine is indeed universal to mankind.

that true existence...."³⁴ And Meister Eckhart observes: "I have no doubt of this, that if the soul had the remotest notion of what Being means, she would never waver from it for an instant."³⁵

A corresponding distinction needs of course to be made when it comes to the idea of "the whole." Here too it is imperative to distinguish between a cosmic and a supra-cosmic or divine "wholeness," which needless to say are different to the point of being incomparable. Even the cosmos taken in its entirety is as nothing compared to the divine plenum, the greatness of which is literally immeasurable. What are twenty billion light years, say, or anything whatever that is bounded by number, in comparison to the Infinite? And let us note that even the word "comparison" proves to be inappropriate and inapplicable. Where God is, nothing else can stand beside Him to be compared: truly He "dwells in unapproachable light."³⁶

Such is the time-honored doctrine; yet Teilhard thinks otherwise. In defiance not only of Christian tradition, but of the perennial metaphysical wisdom of mankind, he opts in favor of holistic pantheism. The notion of "wholeness" evidently inspires him: "Ultimately, our thought cannot grasp anything but the Whole," we are told, "nor, when it really comes to the point, can our dreams entertain anything but the Whole."³⁷ But the question is, *which* Whole: the cosmic, or the divine? And if there be any doubt on that score, the following passage will put that doubt to rest:

> Thus, from the patient, prosaic, but cumulative work of scientists of all types, there has spontaneously emerged the most impressive revelation of the Whole that could possibly be conceived. What the ancient poets, philosophers and mystics had glimpsed or discovered (primarily by intuition), what modern philosophy demands, more rigorously, in the order of metaphysics, science of today has brought within our grasp....³⁸

---

34. *In Joannis Evangelium*, 38.10.
35. *Meister Eckhart* (C. de B. Evans, trans., London, SPCK, 1940), vol. I, p. 206.
36. 1 Timothy 6:16.
37. CE, p. 58.
38. Ibid., p. 63.

Only it needs to be said that this is actually *not* what the wise men of old "had glimpsed or discovered"—not, at any rate, so long as one pays these "poets, philosophers and mystics" the courtesy of taking them at their word. To be sure, God "monopolizes in himself the totality of being"; and this is precisely the import of the Biblical *ego sum qui sum*.[39] But that "being" is absolute: it is "pure being" one could also say: Being that is not subject to conditions or bounds. And as all "the ancient poets, philosophers and mystics" have testified, that Being cannot be grasped: it eludes our cognitive grasp by the very fact of being unconditioned. God, then, is *not* the cosmic whole: He is infinitely higher, infinitely greater than the universe in its totality, We must not forget that the cosmic whole, however vast it may be, is yet conditioned, yet *bounded*: if it were not, it would not in any sense be "cosmic," nor could it be an object of scientific inquiry. The notion of an "unbounded cosmos" is in fact a contradiction: the cosmos is after all *defined* by its bounds. We need not belabor the point: contrary to Teilhard's ontological surmise, the cosmos as a whole, no less than its parts, remains perforce and forever "a lower or secondary mode of being."

As we have noted earlier, Teilhard's postulated Omega had been earmarked from the start to serve as the god-term of his system. Not that God reveals Himself to us through that—real or imagined—Center, a notion which of course would be perfectly orthodox: after all, the cosmos *is* a theophany, as St. Paul declares: "For the invisible things of Him are clearly seen, being understood by the things that are made, even His eternal power and Godhead."[40] But such is not the teaching of Teilhard de Chardin. In an evolutive universe, let us recall, "God is not conceivable (either structurally or dynamically) except insofar as he coincides with ... the center of convergence of cosmogenesis." Teilhard is adamant on the point: "In future only a God who is func-

---

39. I have dealt with this question in *Cosmos and Transcendence* (Tacoma, WA: Sophia Perennis / Angelico Press, 2008), pp. 44–48.

40. Romans 1:20.

tionally and totally 'Omega' can satisfy us."[41] This is evidently what it takes to transform "the Father-God of two thousand years ago" into an up-to-date "cosmogenesis-God."[42] In Teilhard's system God has become "totally Omega" by an act of definition: "God can only be defined as a *Center of centers*,"[43] we are told.

But why? Why should all other conceptions of Deity be ruled out? After all, our forefathers conceived of God in so many ways! As we learn from Dionysius, they thought of Him as Being, Life and Intelligence; as Wisdom, Reason and Truth; as Power, the Great and the Small; as Peace and as Holiness. There is in fact no end to His "divine names." But one needs also to realize that God is beyond every name and conception, "that while He possesses all the positive attributes of the universe (being their universal Cause), yet in a stricter sense He does not possess them, since He transcends them all."[44]

Strictly speaking, God cannot be "defined": to define, after all, is to limit, to set bounds; but who can "measure" God? How can the Infinite be circumscribed? And that is why "those who cling to the objects of human thought"[45] delude themselves when they theologize, and why only "the poor in spirit" can enter into the Presence of God.

But Teilhard insists not only that God can be defined, but that there is only one valid definition: "God can only be defined as a *Center of centers*." And he goes on to say that "in this complexity lies the perfection of His unity."[46] Teilhard's ever-helpful editor, fearing perhaps that the reader might find this hard to grasp, has supplied the following elucidation: "The more God *is*, the more power He has to center and perfectly personalize. Consequently unchangingness belongs no less to the richness of an infinite complexity supremely unified than to an essential simplicity." Spoken like a true disciple! "The more God *is*": the very phrase takes one's breath away. But then, once God has been *defined* as "a Center of centers," He evidently ceases to be absolute and transcendent and becomes

41. CE, p. 240.
42. Ibid., p. 202.
43. HE, p. 68.
44. *Dionysius the Areopagite* (C. E. Rolt, trans., London, SPCK, 1940), p. 193.
45. Ibid., p. 192.
46. HE, p. 168.

tied to the cosmos: after all, a center—even a "Center of centers"—cannot be conceived apart from whatever it be whose center it is.

Such a "cosmogenesis-God," to be sure, cannot be immutable. As a Center of centers He too is subject to change, He too must "evolve"; and such "unchangingness" as is said to belong to "the richness of an infinite complexity supremely unified" could only come about at the end of the evolutive trajectory, as the culmination of the cosmogenetic process, which is precisely its endpoint, the so-called Point Omega.

Yet "we must be careful to note," Teilhard tells us, "that under this evolutive facet Omega still only reveals *half of itself*."[47] And he goes on to explain:

> While being the last term of its series, it is also *outside all series*. Not only does it crown, but it closes. Otherwise the sum would fall short of itself, in organic contradiction with the whole operation. When, going beyond the elements, we come to speak of the conscious Pole of the world, it is not enough to say that it *emerges* from the rise of consciousness: we must add that from this genesis it has already *emerged*; without which it could neither subjugate into love nor fix in incorruptibility. If by its very nature it did not escape from the time and space which it gathers together, it would not be Omega.[48]

But what does this actually mean? If "the other half" of Omega "has already emerged from this genesis," this would seem to imply that it, too, has evolved. All of Omega, then, would be a product of cosmogenesis. But perhaps Teilhard means to imply that there is a transcendent kernel, as it were, something which has existed from the start and presumably constitutes the very heart of Omega. Is that perhaps the entity which "by its very nature" escapes from time and space? But then in what sense could this term, which is supposed to be "outside all series," be said to have "already emerged from this genesis"? Surely "already emerged" implies a preceding "immersion"; and this would seem to rule out the idea that "the other half"

---

47. PM, p. 270.
48. Ibid., pp. 270–271.

THE GOD OF EVOLUTION 133

of Omega could "by its very nature" escape from time and space.

The problem is that Teilhard wants to have it both ways: he wants God to be not only the Evolver, but the Evolved as well. But is this not asking too much? How can Omega be "outside all series" and "the last term of its series" as well? Teilhard tries to resolve this impasse by partitioning Omega, bisecting it if you will: what is supposed to be the supreme unity is somehow split in half! And not until the end of time—when the cosmogenetic process of creative union has reached its term—shall the twain be joined: only then will God Himself be *whole*! That seems to be the theory, and needless to say, the doctrine is strange. It appears, for example, that the Biblical "*ego sum*" needs now to be modified: God should have said, not "I am," but "I shall be." And would it not also have been more accurate if Christ had declared: "I and my Father *shall be* one"?

Leaving aside the delicate question whether any of this actually makes sense, we are beginning to understand at least that the new theology—if there actually be one!—is quite unlike the old: the transition from "the Father-God of two thousand years ago" (the God of Revelation, namely) to a "cosmogenesis-God"—a God who is "functionally and totally Omega"—proves to be more radical than one might have thought.

The new theology, by its very nature, has to do perforce with the Incarnate Christ. But of course the figure of Christ, too, has undergone change: it has been transformed into a kind of cosmic Christ, an Evolver clothed with the body of the universe. And most importantly: it has become an *evolving* Christ, one who is himself dependent upon the cosmogenetic process. As Teilhard tells us dramatically (in an essay completed just a month before his death):

> It is Christ, in very truth, who saves—but should we not immediately add that, at the same time, it is Christ who is saved by Evolution?[49]

49. HM, p. 92,

One wonders, moreover, how this cosmogenetic Christ, "who is saved by Evolution," may be related to God the Father—or are we naïve, perhaps, even to ask? Has not "the Father-God of two thousand years ago" been cast out as an obsolete and unacceptable remnant of a childish age? It is hard to tell just how much of Christian theology we are permitted to preserve. In fact, is anything of *ad intra* theology—of the Trinitarian doctrine!—actually left at all?

Teilhard himself has little to say on this obviously crucial issue. Rarely—and only, as it were, in passing—does he touch upon the subject; and when he does, he is at pains to treat even this question from an evolutionist point of view. We are told, for example, that the trinitarian nature of God "is manifestly the essential condition of God's inherent capacity to be the personal (and, in spite of the Incarnation, the transcendent) summit of a universe which is in process of personalization."[50] But why is it necessary to postulate the existence of *three* transcendent Persons to account for the production of a single "personal summit of the universe"? Yet be that as it may, the statement is doubtless reassuring to many. It is also one of those occasional nuggets of near-orthodoxy which Teilhard's theological defenders may seize upon to refute the charge of heresy.

When it comes to the writings of Teilhard de Chardin one needs however to remember that he does not always say the same thing. Consider now the following passage, which happens to play a major role in *The Phenomenon of Man*: "If the world is convergent and if Christ occupies its center, then the Christogenesis of St. Paul and St. John is nothing else and nothing less than the extension, both awaited and unhoped for, of that noögenesis in which cosmogenesis—as regards our experience—culminates."[51] In the first place it must be recalled that there is not just one "Christogenesis of St. Paul and St. John," but there are actually two: the eternal begetting of the Son or Word that "was in the beginning," and the human birth of Christ, when that Word "became flesh, and dwelt among us."[52] Which "Christogenesis," then, is Teilhard referring to? It obviously

50. CE, p. 158.
51. PM, p. 297.
52. John 1:14.

could not—by any stretch of the imagination!—be the eternal begetting of the Logos: only a "cosmic Christogenesis," after all, could conceivably be envisioned as an "extension" of the evolutive process. But then, what about the *ad intra* Logos doctrine: does Teilhard, or does he not, envision an eternally-begotten Word? One has ample reason to believe that he does not; and as a matter of fact, Teilhard himself implies as much in a passage which it behooves us to consider well:

> In the first century of the Church, Christianity made its definitive entry into human thought by boldly identifying the Christ of the gospel with the Alexandrian Logos. The logical continuation of the same tactics and the prelude to the same success must be found in the instinct which is now urging the faithful, after two thousand years, to return to the same policy; but this time it must not be with the ordinating principle of the stable Greek kosmos but with the neo-Logos of modern philosophy— the evolutive principle of a universe in movement.[53]

Now, it was of course St. John, the Beloved Disciple, who identified Christ with the Logos of Greek philosophy; and one is struck by the fact that Teilhard evidently conceives of this momentous step— which Christians have always regarded as an inspiration of the Holy Spirit—as a kind of philosophical conjecture. The adjective "Alexandrian" is evidently meant to emphasize the local and time-bound nature of this Johannine teaching, and tie that doctrine to the now-antiquated conception of "the stable Greek kosmos." We are led to believe that St. John was attempting, as best he could, to adapt the religious ideas of nascent Christianity to a somewhat primitive and decidedly pre-scientific *Weltanschauung*. And now that we have, at last, discovered the true contours and nature of the universe, we are not only entitled but indeed obligated to employ what he terms "the same tactics" (i.e., *scientific* means) to achieve a better and presumably definitive formulation of the Christian doctrine. And the message is clear: *Teilhard himself is the new Apostle* authorized to overrule St. John by identifying Christ—not with "the Alexandrian

---

53. CE, pp. 180–181.

Logos"—but with "the neo-Logos of modern philosophy—the evolutive principle of a universe in movement."

We are not concerned, at the moment, with the blatant impiety and indeed impertinence of this arrogation, a matter which will be dealt with in the last chapter. What presently concerns us is the fact that Teilhard has indeed rejected the New Testament Logos doctrine, and in so doing, has cast aside *ad intra* theology as such. Gone is the Christian doctrine of the Holy Trinity, and gone too the concept of Biblical and Magisterial inerrancy—which now leaves him free to do as he will.

What concerns Teilhard almost exclusively is the Incarnate Christ, conceived as the Center towards which the universe evolves: "It was in order that he might become Omega that it was necessary for him, through the travail of his Incarnation, to conquer and animate the universe."[54] It is unclear, of course, what "Incarnation" could possibly mean now that Christ has been identified with "the neo-Logos of modern philosophy." Can it actually be supposed that this "neo-Logos" became Jesus of Nazareth? That notion is evidently too incongruous to be seriously entertained. One is therefore not surprised to learn that "like the Creation (of which it is the visible aspect) the Incarnation is an act co-extensive with the duration of the world."[55] Along with the Holy Trinity, it appears, Teilhard is forced to discard "the historical Christ" as well: his evolutionist premises leave him no choice.

One cannot but be amazed at the temerity of this self-styled prophet. It seems scarcely to trouble him that he flatly contradicts what Scripture has taught and Christians have believed and staked their life upon for two thousand years. Has Teilhard persuaded himself, perhaps, that his credentials as a "scientist" give him *carte blanche* to "readjust the fundamental lines of our Christology,"[56] to put it in his own words?

54. SC, p. 54.
55. Ibid., p. 64.
56. CE, p. 139.

What presently concerns us, however, is his conception of God. Leaving aside all other questions, let us try to understand, as clearly as we can, what it is, exactly, that Teilhard is telling us. Are we to suppose, in particular, that Christ descended into the world (whatever that might mean!) at the beginning of time to clothe Himself in its particles, its plasmas, or its primordial slime? Teilhard does not commit himself on this point. The idea of pre-existence is obviously not to his liking; and yet, if there is no pre-existence and hence no "descent," how can one speak of Incarnation at all? There are passages in the Teilhardian opus which do seem to accede to this logical demand; we are told, for example, that "the Redeemer could penetrate the stuff of the cosmos, could pour himself into the life-blood of the universe, only by first dissolving himself in matter, later to be reborn from it"[57]: to speak of "first dissolving himself," after all, is to imply a pre-existence. Yet elsewhere Teilhard says just the opposite: "If God wished to have Christ," he tells us, "to launch a complete universe and scatter life with a lavish hand was no more than he was obliged to do."[58] In other words, *there can be* no pre-existence, no Christ "before ever the world began": it simply was not within God's power "to have Christ" without first "launching" a universe. So too we read in *Mon Univers:* "God did not will individually (nor could he have constructed as though they were separate bits), the sun, the earth, plants, or Man. He willed his Christ—and in order to have his Christ, he had to create the spiritual world, and man in particular, upon which Christ might germinate."[59] The implication, once again, is that Christ could not have existed prior to and independent of the cosmogenetic process—a far cry from what Teilhard had said when he spoke of Christ as "first dissolving himself in matter, later to be reborn from it." How could Christ "first dissolve himself in matter" if, to "have Christ," God was obliged first to create man—by way of evolution, no less!

Rhetoric aside, there is no place in Teilhard's doctrine for Incarnation in the Christian sense. In his system Incarnation reduces

57. SC, p. 60.
58. CE, p. 32.
59. SC, p. 79.

essentially to cosmogenesis: it turns into the birth of galaxies, of stars and planets and protein molecules. It could not be otherwise; for as Hans Urs von Balthasar has astutely observed, the idea of universal evolution is for Teilhard de Chardin finally "the only category of thought." So long as one assumes that nothing exists which has not "evolved," there obviously *can be* no Incarnation. So too, if "nothing can enter into the universe that does not emerge from it,"[55] it follows that nothing "can enter into the universe" at all: not even Christ! As Teilhard tells us plainly: "For Christ to make his way into the world by any side-road would be incomprehensible." Yes, Incarnation as Christians understand the term *is* incomprehensible to a thoroughgoing evolutionist simply because he assumes from the start that the universe in which we find ourselves *is all that exists*.

To those who believe in the reality of an absolutely transcendent God, on the other hand, the idea of the Christic Incarnation is not incomprehensible in the least. It is in fact what Scripture declares and what Christians have always believed: Christ did "make His way into the world"—not, indeed, by a "side-road," but by a mysterious eschatological path which He was again to retrace at the Ascension, when He returned to His super-celestial abode, clothed in the transfigured flesh of His humanity. And this is in fact the Path by which every Christian is called to ascend *in Christ*, who is Himself the Way and the Gate by which the faithful enter into the Kingdom of God.[60]

For Teilhard, on the other hand, "nothing can enter into the universe that does not emerge from it": there *is* no Way, thus, no Gate leading into or out of this universe. What he falsely terms "the Incarnation" can therefore be nothing more, ultimately, than an "ascent from plurality," a process which he conceives to be co-extensive with the history of the universe. "That is why the Incarnation of the Word was infinitely painful and mortifying, so much so that it can be symbolized by a cross."[61]

---

60. John 14:6, 10:9.
61. SC, p. 60.

Having virtually annihilated Christology by his "readjustment" of its "fundamental lines," Teilhard apparently senses the need for some damage control. If the old notions do not fit into his system, he can alter them, "cut them down to size" till they do. Even Jesus of Nazareth, it seems, has not become altogether obsolete: a place can be found for him as well. Although "creation, incarnation and redemption are not facts which can be *localized* at a given point of time and space," Teilhard assures us that "it is nevertheless true that all three can take the form of particular *expressive* facts.... These historical facts, however, are only a specially heightened expression of a process which is 'cosmic' in its dimensions."[62] Thus, in Teilhard's eyes, the birth of the Redeemer is but "a specially heightened expression" of a universal cosmic process that goes on, always and everywhere, throughout the universe. But in that case, one might well ask, what is it that distinguishes the "historical" Christ? If it be true that Jesus of Nazareth came into existence by the same evolutionary path which (according to the Darwinists) has been traversed by us all, why should he be unlike other men? Why should he have been singled out, as it were, "to become Omega"? Was he, perhaps, accidentally endowed with a bigger brain?

Teilhard seems in any case to concur with the judgment of Christianity that by His Resurrection the Incarnate Christ became *kyrios*, Lord of the world: "It marks Christ's effective assumption of his function as the universal center"[63] he maintains. One wonders, of course, how it is possible for an evolutionist to claim that a deceased person could become a universal center. But this is a question which does not seem to interest Teilhard particularly: it would appear that on this point, at least, he is content to invoke the Christian *status quo*. It seems that even Scripture and Tradition have yet their use!

The Resurrection, then, is a fact: "It marks Christ's effective assumption of his function as the universal center." Now this statement (rightly understood) is of course entirely orthodox. But the

62. CE, p. 135.
63. SC, p. 164.

next sentence, already, is not: "Until that time," Teilhard goes on to say, "he was present in all things as a soul that is painfully gathering together its embryonic elements." Not at all! Before His Resurrection Christ was indeed present in all things, but *not*, most assuredly, "as a soul that is painfully gathering together its embryonic elements" (whatever this less-than-felicitous expression might mean). As the Logos, the eternal Word of God, Christ has always been "present in all things," that is to say, *immanent*; this is not something that came to pass at a particular moment of cosmic history. The immanence of God is a metaphysical fact which, most assuredly, does not coincide with the Incarnation. Nor does it have anything to do with a pre-human "soul," let alone with a painful "gathering together of embryonic elements." Christianity teaches that Christ assumed a soul when He assumed a body: and this happened when the Blessed Virgin conceived—and not a moment before! And so, too, the Word became—not a plasma, or a fish, or a reptile—but indeed a *man*. And this alone, let us note, implies that the Incarnation is *not* "co-extensive with the duration of the world"; for no one, least of all a Darwinist, would maintain that mankind existed on earth ever since the world began.

Teilhard conceives of Christ as "the leading shoot" of the Darwinist Tree, the central shoot whose roots extend back "into the furthest limits of the past"[64] whence all things supposedly arise. We are to think of Him as having concentrated within Himself the primordial sap of the universe, the very sap which flows also in our veins. He has gathered together in His own consciousness "the whole mass of passions, of anticipations, of fears, of sufferings, of happiness, of which each man represents one drop." Now this vision—which is not without its strong poetic and mystical appeal—is doubtless correct so far as it goes. It is orthodox Christian doctrine that Christ does recapitulate within Himself "the hopes and fears of all the world." In a sense He does contain each and every one of us within His Soul, in that Sacred Heart said to burn in boundless love. And it is likewise true that His human roots trace back to the very beginning of our race. But we must not forget that Christ has a "double

---

64. Ibid., p. 61.

parentage": He is both Son of God *and* Son of Man. And this is evidently the crucial fact which distinguishes the Incarnate Christ from all other men: Jesus alone is the Son of God, and He alone can say: "I and my Father are one." This is what Christianity teaches, and it is this confession, precisely, that defines a Christian.

But how on earth can this fundamental dogma be interpreted in evolutionist terms? It has always been recognized that God is indeed "the Center of centers"; but only Teilhard de Chardin had ever imagined that God could actually be *defined* as such.

Along with the absolute transcendence of God Teilhard has lost the idea of divine immanence as well. Necessarily so! For as C.E. Rolt observes in his commentary on Dionysius: "The Godhead's Transcendence and Immanence are ultimately the same fact."[65] But whereas transcendence has disappeared, immanence has been confounded with Incarnation.

Teilhard seems to be under the impression that God cannot be immanent in the universe without being "incarnated" therein. He claims in fact that "to create is for God to unite himself to his work, that is to say in one way or another to involve himself in the world by incarnation."[66] This is presumably what it means for God to become "endomorphized" or "Christified." And this "endomorphizing" is supposedly an ongoing process: God is becoming ever more "united to his work," ever more immanent and incarnate in the universe.

What Christianity teaches, on the other hand, is that God has been, from the start, fully immanent in creation. The cosmos has never been without its Center, its "point of contact" with God. And as the eternal Logos, Christ has ever presided over the universe, which was created *in Him, through Him,* and *unto Him* as St. Paul declares.[67] But we must also bear in mind that there has been a

---

65. *Dionysius the Areopagite*, op. cit., p. 184n.
66. CE, p. 182.
67. Colossians 1:16. Presumably *en autoi, di autou,* and *eisauton* refer to the paradigmatic, the efficient, and the final cause of the universe, respectively, the point being that the Logos is all three.

certain falling away from that Center: there has been a Rebellion and a Fall, as we learn from the Book of Genesis. The original harmony and coherence of creation have become compromised. Not that the primordial Center has disappeared: it is there exactly as before. It is *we* who have distanced ourselves from that Center, as it were. In the language of the Old Testament, mankind has departed from the Garden of Eden, and from the Tree of Life which stands in its midst.

Meanwhile the cosmic Center remains immanent in all that exists. But though the Center resides "within," the creature finds itself far removed from that Center: what else is the Fall but an estrangement from God? God is ever present, but *we* are far away: "Thou art within, but I was without" laments St. Augustine.

In terms of this inherently Biblical symbolism we can begin to understand what Christ has accomplished by His Incarnation. One could perhaps put it this way: the Center has swelled, has begun to break into the cosmos as it were, to retake possession of the world. We have strayed, and that Center is now pursuing us to the ends of the earth. And this expansion, this pursuit, began when a Virgin conceived and bore a Child: "and they shall call his name Emmanuel," *God with us*.[68]

God had always been *within*, as we have said; but He has not always been *with us*. This is what happened at Bethlehem: "The Word was made flesh and dwelt among us."[69] Not that the Word came to dwell within the flesh as in a tabernacle: for indeed, as the universal Center Christ had ever been present within all things. No: the Word did not simply *enter* flesh: it *became* flesh as St. John declares. That is the Mystery. And that is why the body of the Incarnate Christ is unlike any other body, and why His actions have a universal significance and an unlimited efficacy. What distinguishes the man Jesus from all other men is quite simply that Jesus *is* God: both man *and* God, to be precise.

This (as we have indicated before) is the crucial point of which we must never lose sight: everything hinges upon this supreme

---

68. Matthew 1:23.
69. John 1:14.

Christological fact. It is the reason why certain events which, historically speaking, took place two thousand years ago, have transformed the world, and why "in their supreme efficacy they continue to be the principle that acts in all human events" as Emile Mersch has said so well.[70] Christ has always been the Center of the universe by virtue of His divinity; but it is by His Incarnation that He became the Head of a transfigured humanity. One should not forget, however, that "if the Saviour is Head through His humanity and in His humanity, He is such only by reason of His divinity."[71]

The Word "became flesh." Let it be said, however, that by no stretch of the imagination does Christianity envisage this act as an evolutive process. First of all it conceives of this Act as a gratuitous intervention on the part of God: it is Christ who, of His own free will, took up human flesh, as it were, in order to breathe His Spirit into that flesh and transfigure it. And though, historically speaking, this redemptive and deifying Act was accomplished in stages, beginning with the Annunciation and ending with the birth of the Saviour, it does not antedate these narrow bounds, much less does it require so many more millennia to complete: by the time the shepherds reached the crib, the Miracle was accomplished. What takes time is not the birth of the Savior, but the resultant transformation of the world.

It is true that St. Paul envisions the Risen Christ as the Center of creation: but not in the sense that a new Center has been formed, or that a hitherto vacant Apex has been occupied. Not at all! By His Ascension Christ as the Incarnate Lord assumed a position and a prerogative which He had always held by virtue of His divinity. And for this reason, too, there can be no dispute as to "which Christ" St. Paul is referring to when he declares that "in Him all things consist"[72]: whether this refers to the eternal Word or to the Risen Lord. One can say that it refers to both: for the two are now forever joined. As Mersch observes: "It is clear that according to the Apostle, Christ's primacy as Head is the continuation of His primacy as

---

70. *The Whole Christ* (Milwaukee, WI: Bruce, 1938), p. 125.
71. Ibid., p. 127.
72. Colossians 1:17.

Word, and that these two prerogatives of excellence mutually explain each other."[73]

Let us also point out that the entire cycle of "sacred history" ending with the Ascension of Christ does not entail the slightest change or transformation in the nature of God. The world does *not* "endomorphize" God, as Teilhard imagines. In His divine nature Christ has suffered no change: neither increase nor diminution can be predicated of the Logos as such. Whatever changes have taken place pertain precisely to His *human* nature. And let us be careful to add that from the moment of the Resurrection that human nature, too, has passed beyond all change. Having accomplished the Passover, it changes no more. As the Church teaches: "He is seated at the right hand of God the Father Almighty"—an image which evidently conveys a sense of stasis, of immunity from change.

What has been transformed and has passed beyond all variability is the human nature of Christ, which has now become eternally united to God. And that is indeed the purpose of the Incarnation: *God became man in order that man might become God,* to quote the great Patristic formula. *In Christo* all men can "become God." But we must understand that both halves of the formula are elliptical. In the first place, God did not "become man" in the sense of suffering transformation, as we have said before. His "taking flesh" is an act of giving, of bestowing Himself upon a lesser nature without suffering the slightest diminution, let alone "increase." So too we need to realize that "becoming God" does not refer to a change of nature, but to a reception, through grace, of what Scripture terms "the glory" of God. What is offered is an eternal participation in the Trinitarian Life—but not identity with God.

His Immanence and Incarnation notwithstanding, the God of Christianity remains the Transcendent, the Incomprehensible, the Absolute which brooks no familiarity. Nothing can bridge or narrow this abyss; and indeed the nearer a saint draws to God, the

---

73. Op. cit., p. 141.

more acutely he stands in awe before the One who "dwells in unapproachable light."[74]

By the same token it can be said that the faith of a Christian is not founded upon ratiocination: it is worlds removed from a "philosophical" conviction. A Christian does not believe in God because the universe converges supposedly to a Center, or because Someone is needed to drive the evolutive process. Presumably he does not hold any such cosmological convictions at all. And if perchance he did, he would not dream of basing his religious beliefs upon such tenets. Christianity teaches that Faith is a divine gift, to be received through an act of humility: no matter how brilliant one may be, what is needed is a certain "poverty of spirit," to use a Gospel term. It could perhaps be said that we believe in the God of Christianity to the extent that we have ceased to believe in anything else. God cannot be simply one certainty among many: He must ultimately be *the only certainty*.

But clearly Teilhard de Chardin thinks otherwise, as he himself admits in a most illuminating passage:

> If as a result of some interior revolution, I were to lose in succession my faith in Christ, my faith in a personal God, and my faith in spirit, I feel that I should continue to *believe* invincibly *in the world*. The world (its value, its infallibility and its goodness)—that, when all is said and done, is the first, the last, and the only thing in which I believe.[75]

It appears that Teilhard wants only as much of God as he can fit into his evolutionist convictions. The Incarnate Christ is supposedly needed as a Center of cosmic attraction, a kind of theistic black hole in which everything will eventually be swallowed up. As Point Omega Christ is there to drive the wheels of progress and consummate the evolutive ascent. Teilhard envisions a Christ who fronts the cosmos, but not the Christ who said "Before Abraham was, I am"[76]: a "cosmic Christ," but not the eternal Logos, not the Second Person

---

74. 1 Timothy 6:16.
75. CE, p. 99.
76. John 8:58.

of the Holy Trinity. Teilhard does not know that notwithstanding what he triumphantly terms "the discovery of Time and Space," the cosmos in its entirety is as nothing before God, and that even the Pantocrator Himself—if for an instant He could be separated from the eternal Logos—would be as nothing too.

Teilhard claims to have found a cosmic use for the Incarnate Christ; but despite certain expressions of theological courtesy, it is plain that he has no need for God the Father. He believes "invincibly" in the goodness of the world, forgetting that "there is none good but God."[77] Teilhard apparently lacks the *sine qua non* of Christian faith: a naturalistic and utilitarian rationalism stands evidently in the way of authentic Christian belief. His is a different faith, as he himself admits. Despite many a Christian sentiment and pious phrase, it appears that in his thought and in his heart the world has gained precedence over God.

It is therefore hardly surprising that Teilhard's theology is not only truncated but secularized: the little that remains savors of the world. His ambition is to make God coincident with the world conceived in evolutionist terms. "When all is said and done," Teilhard's trumpeted "God of Evolution" was presumably indeed "the first, the last, the only thing in which I believe."

---

77. Matthew 19:17.

7

BIBLICAL FALL AND EVOLUTIONIST ASCENT

CHRISTIANITY SPEAKS OF a primordial perfection and a subsequent Fall; and it is hardly surprising that this Biblical doctrine has not been warmly received by Darwinists. Teilhard has surely a point when he observes that "the principal obstacle encountered by orthodox thinkers when they try to accommodate the *revealed* historical picture of human origins to the present scientific evidence is the traditional notion of original sin."[1] What actually militates against the orthodox tenet is not however "the present scientific evidence," but simply the evolutionist credo, for which there is in truth no evidence at all.

We must remember, too, that the Biblical account admits of numerous interpretations, ranging from the more or less "historical" to the profoundly metaphysical. Not that Scripture is ambiguous or equivocal in the ordinary sense, or that theological tradition has not yet been able to ascertain the true meaning of the Biblical text: a single Truth is reflected on various existential planes, and this both explains and justifies the multiple lines of interpretation.

What displeases Teilhard, among other things, is what he terms "the jealous maintenance, as a dogma, of strict monogenism (first one man, and then one man and one woman), which it is in actual fact impossible for science to accept."[2] Now it is questionable, first of all, whether the Biblical "monogenism" can be legitimately interpreted as an "actual fact," by which Teilhard obviously understands

1. CE, p. 36.
2. Ibid.

one that can be verified by scientific means. There *are* other kinds of facts, to be sure, unless it be assumed from the start that theology and metaphysics are void of sense. But even if one should conceive of Adam and Eve as people more or less like ourselves, who lived so many thousand or million years ago in a valley near the Euphrates, the resultant picture of human origins would still not be in conflict with any "actual fact." When it comes to the historical origin of man—or for that matter, of any other species—we find ourselves, scientifically speaking, in a realm of conjecture, of untested and indeed untestable hypotheses. There are no "instruments" to register the advent of a species; and as Teilhard himself admits by his so-called "law of automatic suppression," it appears that not even a fossil record of the earliest progenitors can be found.

This line of criticism, then, leads nowhere—and Teilhard knows it. In an essay entitled "Fall, Redemption, and Geocentrism," for example, he veers away from paleontological considerations almost immediately to talk instead about what he terms "the collapse of geocentrism."[3] In his eyes that stipulated "collapse" militates against the Biblical anthropology: "The fact was that in consequence the seeds of decomposition had been introduced into the whole of the Genesis theory of the Fall."[4] But whatever the historical connection between geocentrism and Darwinist anthropology may be, *logical* connection there is none: it is a very long way from Copernicus to Darwin. What is more, in light of Einsteinian physics itself one cannot claim that Copernicus was right and Ptolemy wrong: the most one may say is that a heliocentric coordinate system yields simpler laws of planetary motion (which, by the way, is precisely what Copernicus himself maintained). What has actually collapsed is not in fact geocentrism

---

3. Contrary to what we are generally led to believe, geocentrism (unlike the "flat Earth" hypothesis, with which it is sometimes compared) is by no means a dead issue. Rarely are we told, for example, that in light of the Michelson-Morley experiment, all of classical physics stands in fact on the side of that "disproved" tenet. The choice actually lies between geocentrism and Einsteinian relativity. And on this question, it seems, the final verdict is not yet in. For an overview of this subject I refer to my chapter on "The Status of Geocentrism" in *The Wisdom of Ancient Cosmology* (Washington, DC: Foundation of Traditional Wisdom, 2003).

4. CE, p. 37.

but rather the capacity of men and women endowed with a college education to perceive more in Nature than *res extensae* and measurable magnitudes. What has well nigh vanished, in other words, is our ability to read the cosmic icon; and it is this disability—rather than the alleged collapse of geocentrism—that has "introduced the seeds of decomposition" not only "into the whole of the Genesis theory of the Fall," but into the Biblical world-view in its entirety: from Genesis right through the New Testament. But this is a question which we shall need to examine at greater length in the following chapter.

It is a fundamental tenet of Christianity that Adam's Fall has afflicted the whole of mankind with the condition of Original Sin: and Teilhard finds this hard to accept. The dogma clearly stands in his way and needs to be somehow neutralized. We are told, to begin with, that "it is impossible to universalize the first Adam without destroying his individuality"[5]: so long as we conceive of him as a *bona fide* person, in other words, Adam could not have had an effect upon the whole of mankind. Such is the first step, the premise of his argument; and the conclusion is this: "Strictly speaking, there is no first Adam. The name disguises a universal and unbreakable law of reversion or perversion—the price that has to be paid for progress."[6]

But before getting too preoccupied with "the price that has to be paid for progress," it might be well to ask what all this entails with reference to the Incarnate Christ: would not the *second* Adam suffer the same disability as the first? If indeed it is "impossible to universalize the first Adam without destroying his individuality," would not the same hold true for the Redeemer? There is after all a connection: "For as by one man's disobedience many were made sinners, so by the obedience of one shall many be made righteous."[7] If the first claim is inconceivable, would not the second be inconceivable too? If "strictly speaking, there is no first Adam," would it not

---

5. Ibid., p. 39.
6. Ibid., p. 41.
7. Romans 5:19.

follow by the same logic that "strictly speaking" there is no second Adam as well?

Obviously even Teilhard cannot go that far: the step would be fatal to his theory. He still has need of the second Adam, who is destined, after all, to become Point Omega. Unlike the first Adam, the second needs therefore to be saved. "The case of the second Adam is completely different," so begins the argument; for whereas the second is situated supposedly at Point Omega, "there is, it is clear, no lower center of divergence in the universe at which we could place the first Adam."[8] But though indeed a "lower center of divergence" may not exist, one has reason to believe that neither does a "higher center of convergence": for as we have come to see, the Teilhardian Omega turns out to be fictitious.

The decisive point, however, is that one cannot legitimately speak of either Adam in scientific terms: the Biblical account simply does not submit to interpretation on the level of scientific discourse. If indeed "the first man Adam was made a living soul; the last Adam a quickening spirit,"[9] what on earth has science to say concerning such things? But Teilhard assumes that it can, and in so doing creates problems which in reality do not exist. The underlying fallacy is to suppose that whatsoever eludes the net of science is *ipso facto* unreal. It is this scientistic postulate that leads Teilhard to conclude that "it is impossible to universalize the first Adam without destroying his individuality," and to imagine that Christianity will collapse unless there be an "Omega Point" to serve as a platform for Christ.

Yet, oddly enough, Teilhard himself admits that the Biblical and the scientific world-views correspond to different levels of vision: "We cannot retain both pictures without moving alternately from one to the other," he observes. "Their association clashes, it rings false. In combining them on one and the same plane we are certainly victims of an error in perspective."[10] How very true! What Teilhard means, however, is something quite different: by situating "the Biblical and the scientific world-views" on different "planes" he

---

8. CE, p. 41.
9. 1 Corinthians 15:45.
10. CE, p. 47.

means to imply that the former belongs in fact to the realm of fantasy. He is telling us, in particular, that the Biblical creation narrative is at best a myth which needs now to be "demythologized," that is to say, transposed into a scientific key. And that is of course what his own theological speculations are intended to accomplish. It is the reason why Teilhard does not rest content until he has convinced himself, six pages later, that Creation, Fall, Incarnation and Redemption—"all four of these events"—are somehow "co-extensive with the duration and totality of the world." Apparently he does not consider it "an error in perspective" to interpret this sacred teaching—"all four of these events"—in alien terms: when it comes to his own fusion of Biblical and scientific conceptions he seems no longer to feel that such a synthesis "clashes" or "rings false."

But let us get back to Teilhard's polemics against the traditional notions of Paradise, the first Adam, and Original Sin. Teilhard is aware, of course, that these Biblical terms can be understood in various ways, although he apparently views the resultant interpretations as so many rival theories, which in truth they are not. There is first of all the "literal" interpretation which situates the Garden of Eden upon this Earth and regards Adam as having been endowed with a body more or less like our own. And as we have seen, Teilhard objects to this view because it is tantamount to a monogenism which can supposedly be ruled out on scientific grounds. Having thus disposed of this particular option, as he seems to believe, he goes on to consider certain versions of what he terms the "Alexandrian explanation." Broadly speaking, these are allegorical interpretations which regard Paradise as a "higher state" and Adam prior to the Fall as someone incomparably more spiritual than we, someone endowed with virtually god-like faculties, which we no longer possess, and a body that differs markedly from our own. And then came the Fall. It hardly matters whether we conceive of this prehistoric catastrophe as an internal disintegration, or as a descent into a lower world and a concomitant decline. The point, in either case, is that the Fall of Adam entails an effective loss of the aforesaid endowments. Let us understand it well: Christianity teaches that this primordial catastrophe—and *not* a Darwinist ascent!—is responsible for the human condition as we know it today.

No wonder Teilhard is displeased. The doctrine is obviously a thorn in his side, a "theory" which needs at all cost to be disproved or in some other way discredited. But how is this to be done? What precisely are Teilhard's arguments? Let us consider a late essay in which he summarizes his objections in the form of three concise statements.

His first point is that "the whole of the extra-cosmic part of the story has 'an arbitrary and fanciful' ring. It takes us into the realm of pure imagination."[11] One might of course ask oneself: who is he to cast the first stone when it comes to "pure imagination"! But to get to his point: what Teilhard apparently fails to realize is that *any* affirmation regarding ultimate realities—be they "first" or "last"—cannot but appear "imaginary" by the very fact that they *are* ultimate, and consequently not subject to the conditions of our phenomenal world.

"Secondly, and much more seriously," he tells us next, "the *instantaneous* creation of the first Adam seems to me an incomprehensible type of operation—unless the word simply covers the absence of any attempt at explanation." Now this is a question we have already examined at considerable length and disposed of in Chapter 4.[12]

Which brings us to Teilhard's third point: "Finally, if we accept the hypothesis of a *single, perfect* being put to the test *on only one* occasion, the likelihood of the Fall is so slight that one can only regard the Creator as having been extremely unlucky." But in fact one can only regard this statement as exceedingly inept: for it makes no sense whatsoever to speak of "likelihood" with reference to a unique event, one that can be put to the test only once. Mathematically speaking, the claim is absurd.

And these are the three arguments, on the strength of which Teilhard has taken it upon himself to impugn two thousand years of Christian tradition!

---

11. CE, p. 193.
12. See pp. 84–89.

The fact is that a primordial state of perfection has no place in an evolutionist world-view: the Darwinist premises do not allow such a state. One way or another, therefore, Teilhard was obliged to rid himself of the first Adam and of the Edenic world which had been his original habitat. In the end we are told that "Adam and Eve are images of mankind pressing on towards God"[13]: this banality is all that remains.

But what about Original Sin—is this to be jettisoned as well? One has the impression that Teilhard would like to very much, but cannot because this would also knock out the idea of Redemption, without which even his "revised" Christianity would make no sense. Yet he does the next best thing: instead of discarding the concept of Original Sin, he recasts it in strictly evolutionist terms, which suffices to remove its sting.

"Whenever we try intellectually and vitally to assimilate Christianity with all our modern soul," Teilhard tells us, "the first obstacles we meet always derive from original sin."[14] That regrettable notion "clips the wings of hope" and "drags us back inexorably into the *overpowering* darkness of reparation and expiation."[15] The fault, however, rests not with the dogma as such, but with the antiquated form in which it has been expressed, which "represents a survival of obsolete static views into our now evolutionary way of thinking. Fundamentally, in fact, the idea of Fall is no more than an attempt to explain evil in a fixed universe."[16]

Let us then consider what becomes of Original Sin once we have become liberated from "obsolete static views." In an evolutionary cosmos, a universe in process of "creative union," evil is to be conceived, basically, as a certain resistance to unification on the part of "the multiple." As Teilhard explains: "Since its gradual unification entails a multitude of tentative probings," the multiple "cannot

---

13. CE, p. 52.
14. Ibid., p. 79.
15. Ibid., pp. 79–80.
16. Ibid., p. 80.

escape (from the moment it ceases to be 'nothing') being permeated by suffering and error."[17] Evil, then, is tantamount to disorder, and disorder is something unavoidable. It is "absolutely inevitable that local disorders appear," and that "from level to level, collective states of disorder result from these elementary disorders (because of the organically interwoven nature of the cosmic stuff)." And that brings Teilhard finally to the subject of Sin, which proves not to be after all Original: "Above the level of life this entails suffering, and, starting with man, it becomes sin."[18]

There it is. But even a cursory look at the new doctrine should give us pause. Why, first of all, should disorder turn into sin on the human level? Why speak of "sin" rather than of disease, error or imperfection? And surely there *is* a difference! The very fact that there is no such thing as sin below the level of man should suffice to make this clear. The point, of course, is that sin presupposes the idea of responsibility: of a free agent who can distinguish between right and wrong.

There may indeed be a connection between sin and disorder inasmuch as sinful acts tend to engender disorderly states. Yet what ultimately counts, from an ethical point of view, is not the physical effect of an action, say the "disorder" to which it may give rise, but the *intention behind the act*: the same physical action may be good, bad or indifferent, depending on what it is that moved the doer to do what he did.

In Teilhard's theory, on the other hand, evil has in effect been identified with disorder and thereby reduced to a thermodynamic quantity: to entropy in fact. But whereas evil may indeed be a cause of disorder—even as "death is the wages of sin"—the two are by no means the same. Death as such, then, which is evidently a falling into disorder on the part of the body, is not a sin, nor is it an evil as St. Paul likewise teaches when he exclaims: "Oh Death, where is thy sting!" In a word, by transposing the conceptions of sin and of evil to a scientific plane—to the level of statistical mechanics, no less!—Teilhard has robbed these terms of their authentic sense. In his

17. Ibid., p. 195.
18. Ibid.

eagerness to reduce the essentials of Christian teaching to the preconceived categories of evolutionist dogma, he has laid the foundations of a scientistic pseudo-theology which not only falsifies the Christian doctrine, but at the same time opens the door to all kinds of monstrous possibilities. It is not by accident that Teilhard advocates various technological interventions to promote "anthropogenesis" and "cerebralization"—not excluding mandatory surgery!

But these are matters that need to be considered in their proper place. Now that we have exposed the premises of Teilhard's brand new "ethics," if one may still call it that, let us see how he justifies that particular "readjustment." He begins by taking his stand on supposedly scientific ground: "The evidence of science is necessarily, and always will be, respected, since the experiential background of dogma coincides with that of evolution."[19] The implication, once again, is that the traditional view does *not* respect "the evidence of science," which is to say that there exist scientific findings which disqualify that view. But this is preposterous: what *are* these findings? And what could they be if not the "facts" hypothesized by Darwinist theoreticians? No wonder Teilhard has elected to make his point, not by adducing actual "evidence of science," but by innuendo alone.

Apart from "respecting" the findings of science the new ethics has supposedly an additional advantage: we are told that "the problem of evil disappears." True enough. It is in fact a modest claim: for as we have already seen, it turns out that not only the problem, but *the very idea of evil* has disappeared. But let us follow Teilhard's argument: "In this picture physical suffering and moral transgression are inevitably introduced into the world not because of some deficiency in the creative act, but by the very structure of participated being: in other words, they are introduced as the *statistically inevitable by-product* of the unification of the multiple. In consequence they contradict neither the power of God nor his goodness."[20] Now, to begin with, if moral transgression is inevitable, if it comes about by force of "statistical necessity," how then can one speak of *sin*? Or to put it another way: the very concept of moral transgression

---

19. Ibid., p. 196.
20. Ibid.

presupposes the idea of human freedom: the possibility, in other words, of *not* transgressing. But perhaps this too has lost its validity: perhaps, in a Darwinist universe, the concept of morality no longer presupposes the postulate of human freedom. But even if one were to redefine morality along Darwinist lines as the use of one's brain to minimize disorder, evil (now conceived as disorder) remains, and so consequently does the so-called "problem of evil." Teilhard misleads us when he pretends to resolve that problem by claiming that "statistically inevitable" disorders "contradict neither the power of God nor his goodness." Yes, such is the case so long as one conceives "the power of God" as falling short of omnipotence: but in that case there never was a "problem of evil" in the first place. So far from solving that problem, Teilhard has sidestepped the issue by implicitly denying one of its premises: i.e., the omnipotence of God.

Teilhard fails to grasp that evil is caused, not by a natural processes, but by a personal agent: it is man as distinguished from Nature who stands at fault. To put it in metaphysical terms: evil derives, not from the material, but from the spiritual pole of creation. Strange as it may seem at first glance, man's capacity to sin derives from the *spiritual* side of his nature: from the intellect or rational faculty, namely, which distinguishes man from animals. It is the abuse of this inherently divine power that gives rise to sin.

In seeking the origin of evil in the operations of Nature—the alleged resistance of "the multiple" to unification—Teilhard is looking in the wrong direction. Whatever be the resistance or opposition by which evil enters into the world, it is a *willful* resistance, a *willful* opposition. What evil resists or opposes, moreover, is not the creative act—which nothing created *can* oppose—but the created order. What it attacks is inherently the primordial world, the pristine universe as it issues from the hand of God.

We need to recall, once again, that the world was not created *in* time, but *with* time, in that supra-temporal "beginning" to which Genesis 1:1 alludes: that is to say, it was created "instantaneously." And this alone renders the creative act irresistible: "He spoke, and it was done; He commanded, and it stood fast." God creates, not one by one in temporal sequence, as some imagine, but *omnia simul*, all at once.[21] Who, then, could oppose this Act?

It follows by the same token that the creation as such is perfect: if nothing can impede or thwart the divine creative will, how could anything be amiss? And Scripture confirms this when it declares that God beheld the creation and saw that it was "good."

What then has happened: what has gone wrong? For it hardly needs pointing out that the world in which we find ourselves falls short of perfection, and that we ourselves are imperfect as well. It follows that there *must* have been a Fall!

To get to the heart of the matter, we need to recall that evil is caused, not by the operations of Nature, but by the intervention of personal agents. The great fact—so often overlooked—is that God created not only natures or "things," but *persons* as well. Yet one cannot say "what" a person is; for as Richard of St. Victor astutely points out, a person is actually not a "what?" but a "who?"

Theology teaches that the mystery of personhood is rooted in God. Because *God* is a Person—or three Persons, to be exact—man can be a person too: "And God said, Let us make man in our image, after our likeness."[22] It is to be noted that here God speaks of Himself in the plural, as if to imply that the image is of that in God which *is* plural: the divine *hypostases* or Persons, namely, as distinguished from the Godhead.

There is something godlike, then, in personhood, or better said, in a person, whoever he or she may be. Only one needs to bear in mind that the person is not in fact the human individual, as we commonly suppose. Individuality, one might say, is no more than a mask woven, as it were, of attributes which belong in truth, not to the person as such, but to his nature. And a nature is something possessed by many beings: it pertains to a species as the Scholastics would say. But a person is unique: "one of a kind" if you will; only in this instance one cannot speak of "a kind" at all. We sense that uniqueness first of all in ourselves: it almost seems to each of us that "I am

---

21. Ecclesiasticus 18:1.
22. Genesis 1:26.

the only I" in the universe. But at times we sense that same uniqueness in another as well; and it is at such moments that we begin to *know* that person. Yet we never quite succeed: the person himself remains an enigma, a profound mystery—as befits an image of God.

Now it might seem strange that evil should arise precisely from what is indeed most godlike in creation. Yet it actually is not; for as Georges Florovsky observes: "The human fall consists precisely in the fact that man limits himself to himself, that man falls, as it were, in love with himself. And through this concentration on himself man separated himself from God and broke the spiritual and free contact with God. It was a kind of delirium, a self-erotic obsession, a spiritual narcissism."[23] Evil, then, results from a certain "break with God." It stands to reason, however, that a "break" can only take place at "the point of contact" as it were, which in this case is evidently the highest point in man. Nor is it hard to understand that the resultant separation from God should prove to be—not just disastrous—but indeed *fatal.*

What then is the Christian response to Teilhard's "problem of evil"? It hinges clearly on the fact that *personhood implies freedom.* We need to understand, moreover, that this freedom, pertaining as it does to the creature, can only be a certain power derived from God. It is something bestowed upon us by God, and of all His gifts the most precious: for *by means of this God-given freedom we can attain eternal life in God.* This very freedom, however, entails a risk—a dreadful possibility!—for it bestows the option of opposing oneself to God's will. What then is the source of evil: whence does it spring? The Christian answer is crystal clear: evil springs from a willful opposition to God, which is sin.

Now we must realize that God, having once bestowed upon us the incomparable gift of freedom, does not take it back. He does not revoke that freedom, even when it is put to deadly use. Unbelievable as it may seem, it is rather His own power which God relinquishes, as it were, in the face of the rebellious creature. As Vladimir Lossky writes most beautifully:

---

23. *Collected Works*, vol. 3 (Bellmont, MA: Nordland, 1976), p. 85.

God becomes *powerless* before human freedom; He cannot violate it since it flows from His own omnipotence. Certainly man was created by the will of God alone; but he cannot be deified by it alone. A single will for creation, but two for deification. A single will to raise up the image, but two to make the image into a likeness. The love of God for man is so great that it cannot constrain; for there is no love without respect. *Divine will always will submit itself to gropings, to detours, even to revolts of human will* to bring it to a free consent: of such is divine providence, and the classical image of the pedagogue must seem feeble indeed to anyone who has felt God as a beggar of love waiting at the soul's door without ever daring to force it.[24]

What more can one say? This, perhaps, is as close as we can come to resolving the so-called problem of evil; and to anyone "who has felt God as a beggar of love waiting at the soul's door" it will suffice.

It may be enlightening at this point to recall an ancient exegetical tradition approved by St. Augustine, which interprets the story of Adam and Eve in anthropological terms. As Meister Eckhart writes in his *Liber parabolarum Genesis*: "The saints and theologians generally interpret what is written in the third Chapter symbolically (*parabolice*) and understand by the serpent the sensual nature (*sensitivum*), by the woman the lower mind (*inferius rationale*), but by the man the higher reason (*superius rationalis*)."[25] And he goes on, a little later, to explain from this point of view what has happened to man in consequence of Original Sin:

> But after the highest power of the soul had lost its connection to and aspiration towards God (*adhaesione et ordine a deo*)

---

24. *Orthodox Theology* (Crestwood, NY: St. Vladimir's Seminary Press, 1978), p. 85.
25. *Die Lateinischen Werke* (Stuttgart: Kohlhammer, 1965), vol. 1, p. 602. This interpretation, incidentally, throws much needed light on the teachings of St. Paul (so unpopular nowadays!) regarding the rightful position of the woman *vis-à-vis* her husband and the Church, as given in 1 Corinthians 11.

through the commission of sin—"But your iniquities have separated between you and your God"[26]—all the powers of the soul, the lower mental as well as the sensual, were successively cut off from the higher reason and its hegemony....[27]

In other words, the hierarchic structure of the human compound came undone, leaving mankind "sick unto death." As St. Thomas Aquinas likewise explains:

> By the sin of our first parent original justice was taken away, by which not only were the lower parts of the soul held together under the control of reason, without any disorder whatever, but also the whole body was held together in subjection to the soul, without any defect. Therefore, when original justice was forfeited through the sin of our first parent, just as human nature was stricken in the soul by the disorder among the powers, so also it became subject to corruption, by reason of disorder in the body.[28]

The picture that emerges from these elucidations is exceedingly clear. So long as man lived in his original state of innocence he remained near to God and fully integrated. To put it in trichotomous terms: his spirit was united to God, his soul to his spirit, and his body to his soul. Thus united, part to part and all to God, he was *one* being, one theomorphic organism. And then came the Fall by which that unity was broken: man became disunited, fragmented as it were. It was rebellion from beginning to end. First the spirit rebelled against God through "disobedience," which constitutes the Fall properly so called. But this catastrophe gave rise to a second rebellion: for in the wake of the Fall the soul rebelled against the spirit through self-love and concupiscence. And this in turn gave rise to a third: a rebellion of the body, namely, against the soul through inertia if you will: the impotence of matter per se. We need to realize that by this triple revolution a reversal of the natural order

---

26. Isaias 59:2.
27. Op. cit., p. 612.
28. *Summa Theologiae*, I–II, Quest. 85, Art. 5.

has taken place: man as we know him is by no means in his natural state. As Vladimir Lossky comments elsewhere:

> The spirit must find its sustenance in God, must live from God; the soul must feed on the spirit; the body must live from the soul—such was the original ordering of our immortal nature. But turning back from God, the spirit, instead of providing food for the soul, begins to live at the expense of the soul, feeding itself on its substance (what we usually call "spiritual values"); the soul in turn begins to live with the life of the body, and this is the origin of the passions; finally, the body is forced to seek its nourishment outside, in inanimate matter, and in the end comes on death. The human complex finally disintegrates.[29]

It may be of interest to ask whether the human body as generally conceived might not actually correspond to the "coats of skins" which God is said to have made for Adam and Eve following their transgression[30]: we know that St. Gregory of Nyssa and other theologians of high rank have speculated along these lines. Could it be that the true or primordial body still exists beneath that "outer shell," that *annamaya-kośa* or "sheath made of food" as it is called in the Vedantic literature? Could it be that what we normally take to be the body is but an outer shell to be shed at the moment of death, something destined indeed to disintegrate? These are matters, in any case, we are ill-equipped to probe, and on which the Church at large has refused to make definitive pronouncements.

Suffice it to say that something in man has become subject to death; and that something, moreover, is not just the body. We must not say (as did certain Platonists) that the body dies in its entirety while the soul remains fully intact: for it stands beyond doubt that the Fall has affected the soul as well. It too has undergone a profound transformation, it too has become destabilized. In a sense it has become split in two; and it scarcely matters whether we call one

---

29. *The Mystical Theology of the Eastern Church* (Crestwood, NY: St. Vladimir's Seminary Press, 1976), p. 128.
30. Genesis 3:21.

part "real" and dismiss the other as a mere superimposition. The fact remains that something of our inner self has become mortal, and that the catastrophe of death affects not only our body, but our psyche as well.

To put it in Pauline terms: we have become split into an outer and an inner man, a spiritual and a carnal being. And it happens to be the latter which on the whole has waxed in the course of history, and in the mass of humanity has become ever more dominant. The point has been reached where countless men and women around the globe find it almost impossible to realize that they are *more* than a carnal being: our spiritual nature has for the most part become almost totally eclipsed. And this in itself explains why the typical man of our time has no conception of the Fall, and why the notion of Original Sin strikes him as an aberrant fantasy, fabricated by morbid or infantile theologians. It explains why, "whenever we try intellectually and vitally to assimilate Christianity with all our modern soul, the first obstacles we meet always derive from Original Sin." No wonder! Without some awareness, however dim, of human nature in its pristine purity, and without the least conception of the Fall—no inkling even of a higher antecedent—one can indeed have no sense of Original Sin, nor any tolerance for the idea. But let us also realize that these higher perceptions are precisely what our present civilization— more effectively perhaps than any other—is engaged to eradicate.

It could hardly be otherwise. The Fall itself entails a progressive "forgetting," an ongoing loss of recollection. Not only does the collective memory of man's primordial nature grow ever more dim, but the entire spiritual order tends to recede from the human field of vision. Teilhard is surely right when he points out that "there is not the least trace on the horizon, not the smallest scar, to mark the ruins of a golden age or our cutting off from a better world."[31] But he is thoroughly mistaken when he supposes that this in itself renders the Biblical doctrine implausible: for as the preceding reflections make clear, what he cites by way of refutation is exactly what the doctrine itself leads us to expect. The expulsion from Paradise is not a historical fact, an event that could in principle be detected by

---

31. CE, p. 47.

"traces on the horizon of time": it is rather a fall into oblivion, a reduction of the cognitive possibilities effectively open to man. What has happened, basically, is that we have lost our spiritual sight; as St. Paul wrote to the Corinthians: "The natural man [*psychikos anthropos*] receives not the things of the Spirit of God: for they are foolishness unto him: neither can he know them, because they are spiritually discerned."[32]

It must be understood that this carnal or outer man came into existence by way of the Fall. Originally there was only one man; and now there are, as it were, two. The Fall constitutes, as we have seen, a disintegration in the most literal sense. And out of this disintegration there has emerged the post-Edenic man, who may be characterized as the ego-centered man. He is the *anima-corpus* compound, one might say, cut off from its spiritual source. And he is *psychikos anthropos* or "psychic man" inasmuch as his life is centered upon the psyche as distinguished from spirit or *pneuma* in the Pauline sense. But who is this *psychikos anthropos* but the man we are all familiar with! The spiritual or pneumatic man in us has after all become eclipsed. Yes, we do—thank God!—catch a glimpse of him now and then; and yet it is mainly the outer man we know. He alone, moreover, is recognized in our anthropology and glorified by humanists around the globe: he is all "our modern soul" can envision. Seyyed Hossein Nasr is no doubt right: "There has never before been as little knowledge of man, of the *anthropos*."[33]

What we know—and alone *can* know from what Teilhard is pleased to call "the phenomenal point of view, to which I systematically confine myself"—is no more than the outer man, the *psychikos anthropos*. Here the knower is himself the *psychikos anthropos* who "receives not the things of the Spirit of God: for they are foolishness unto him: neither can he know them, because they are spiritually discerned." It is to be noted that St. Paul distinguishes between "receiving" and "knowing": not only are we incapable (on this level) of any direct spiritual knowledge, but we cannot even receive the

---

32. 1 Corinthians 2:14.
33. "Contemporary Man, Between the Rim and the Axis," *Studies in Comparative Religion*, vol. 7 (1973), p. 116.

higher teaching when it is offered to us: not even "as through a glass, darkly" can we behold the truth! In a word, it is the innate tendency of the *psychikos anthropos* to be profane and carnal in his *Weltanschauung*. No wonder Teilhard—who exults in what he terms "our modern soul"—should find it imperative to reinterpret and fundamentally recast the spiritual teachings of Christianity!

Yet there is still more to be said; for it is vital to understand that the Fall is not simply a pre-historic event, but is also in a sense an ongoing process: a daily occurrence almost. As Frithjof Schuon observes: "This drama is always repeating itself anew, in collective history as in the life of individuals."[34] Understandably so! Destabilized by the original Fall, man is endowed with an inherited propensity to become ever more alienated from his spiritual source and center. By virtue of Original Sin—which is not after all the invention of theologians!—it is his natural tendency to draw away from that primordial Center. His indigence is thus compounded: "He that hath not, from him shall be taken even that which he hath."[35] One centrifugal step prompts another; and the nearer he draws to the periphery, the more he falls prey to its literally fatal attraction.

Yet to the outer man, the *psychikos anthropos,* this loss appears as gain, this descent as "progress"—right up to the point of catastrophe. As the spiritual man wanes, the carnal man waxes, and extends his dominion over the face of the earth: there is an inverse ratio between the two. No wonder that to the carnal man the pageant of history presents itself on the whole as an ascending trajectory! Blind to the spiritual world, he sees not the treasures he is leaving behind; as with Esau, it pains him not to give up his birthright for a mess of pottage.

So the Fall continues: our expulsion from the Garden of Eden—and from the Tree of Life which stands in its midst—has not yet reached its term. But it continues largely unperceived. What presents itself instead, to the outward-bound observer, is the periphery loom-

---

34. *Light on the Ancient Worlds* (London: Perennial Books, 1965), p. 44.
35. Mark 4:25.

ing ever larger before him: Teilhard's so-called "discovery of Time and Space" is evidently expressive of this fact. Meanwhile the disintegration of the *anthropos* continues at an increasing pace. Originally there was one man, as we have said; and then, following the primordial Fall, there were two. But it happens that the *psychikos anthropos* is himself disintegrating. He has become divided, first of all, into a conscious and an unconscious part of himself, a dichotomy which mirrors the primordial split. But the process of disintegration does not stop at this partition; in the memorable words of George Herbert: "Oh, what a thing is man! How far from power, from settled peace and rest! He is some twentie sev'rall men at least each sev'rall houre."

Who does not know this, who does not recognize it in himself? And who could be so blind as not to see that contemporary civilization thrives on the effects of that dispersion, which it promotes by every conceivable means.

There are also, of course, forces in play which promote human integration on various planes. And above all there are still men and women who live an authentically religious life, a life centered upon God, which integrates not only on a psychic plane, but in varying degrees connects us to the spiritual realm. Religion in general (from *re-ligare*, "binding back") is the great force which opposes the ongoing Fall.[36] Yet as we know very well, religious influence has long been in decline: think of the trajectory from the High Middle Ages, say, through the Renaissance and so-called Enlightenment to modernity, not to speak of postmodernity! However much or little of authentic religion may have survived that onslaught, the fact is that Western civilization has been de-Christianized. No longer constrained—no longer "tied back"—by religious norms which he deems to be outmoded, the proverbial man of our time is free at last to follow his centrifugal bent without let or hindrance, and in so doing to consummate what Christianity terms the Fall.

---

36. This is not the place to enter upon an explanation regarding the uniqueness and primacy of Christianity, which in truth is not merely *a* religion, but *the* religion. As I hope to argue elsewhere, Christianity is its own genus, the genus of religion *per se*.

8

THE NEW ESCHATON IN HISTORICAL PERSPECTIVE

IF THE FRENCH JESUIT has actually inverted the Christian doctrine—if he has deftly turned it upside down—we must not forget that, given the contemporary climate of thought, he did not in fact have very far to go. Teilhard de Chardin is no isolated thinker, and not nearly as unprecedented as some have imagined. He embodies rather a major trend, which indeed he epitomizes: the very movement, namely, that has ushered in the modern age. Let us consider that movement.

Surprisingly, perhaps, the roots of modernity trace back to the great era of Scholasticism. Notwithstanding the unquestioned orthodoxy of its illustrious exponents, it was actually the genius of the thirteenth century that gave birth to what later came to be "rationalism" in the current sense. As some had forewarned, the Scholastic enterprise was not without peril: from the outset there was danger that it could give rise to an over-valuation of human rationality—of discursive thought if you will—which in time might prove fatal to Christian faith.

One must remember that reason, too, has its limitations; it does not simply coincide with intellect, or with intelligence as such. For all its apparent prowess, it constitutes but one particular mode of knowing, a mode which in fact derives from a higher faculty. Discursive thought constitutes after all an indirect way of knowing: a knowing "by reflection" as we say. Thought as such, moreover, belongs evidently to the psychic plane: to the *psychikos anthropos* in fact. We should not be surprised, therefore, that it does not cover the entire ground of knowing, as one is prone to suppose. It happens that the spiritual man in us has cognitive means of his own, and as St. Paul declares, it is only by that higher and indeed god-like faculty that man is able know "the things of the Spirit of God."

Now the great Scholastics, of course, understood this perfectly well. They did not idolize human rationality as we do, nor did they fail to realize that the spiritually awakened man can readily dispense with syllogisms. They knew moreover how to combine reasoning with spiritual contemplation; in their hands logical argument could serve as a catalyst of intellection in the true sense. But unfortunately this spiritual art was not passed on: generally speaking, what has come down to us are the more external and contingent aspects of Scholasticism—the instrument, if you will, the tools once used by the masters. And the very perfection of that instrument may have contributed to the subsequent decline of spiritual vision; the "letter," after all, does "kill."

And as a matter of historical fact, no sooner had the great masters passed from the scene than signs of decadence and disintegration began to appear. In the fourteenth century already a sterile but yet ostensibly Christian rationalism—a rationalism that was fast losing touch with the realities of the spiritual order—begins to come into view, along with rudiments of skepticism and philosophic doubt. No longer was Reason the handmaid of Intellect: of a power in us that needs to be awakened through grace and spiritual art. Instead, Reason declared its own autonomy: and that was the fatal step, the lethal "bite" all over again.

It was above all the birth of rationalism that brought the Middle Ages to a close; and let us understand it well: what followed *was* a fall. As Seyyed Hossein Nasr well observes: "Renaissance man ceased to be the ambivalent man of the Middle Ages, half angel, half man, torn between heaven and earth. Rather, he became wholly man, but now a totally earthbound creature."[1]

It is hardly feasible, within the confines of this chapter, to recount even the principal stages and dimensions of this fateful "evolution"; instead, I would like to draw attention to a salient concept which emerged quite early in the course of that development, and began

---

1. *Man and Nature* (London: Allen & Unwin, 1968), p. 64.

immediately to play a decisive role: the idea, namely, of "progress." In the post-medieval world the concept proves to be essential; our post-Christian civilization demands that idea. Once Heaven was closed and man had become reduced in effect to an earthbound creature, a substitute had to be found, an Ersatz that could somehow take the place of the spiritual Eschaton. Progress then—the specifically modern notion of a collective utopia to be achieved by human industry—replaces by stages and degrees the quest for God, and ultimately becomes confounded with that quest. Whereas the *idea* of progress is initially conceived in secular terms, the veneration of *Progress* blossoms eventually into a mysticism of sorts, a futuristic religion which claims to fulfill and supersede all the religions of the past.

And this is manifestly the point at which Teilhard de Chardin enters upon the scene. What presently concerns us, however, is the fact that once the world has become flattened in the collective imagination—once an effective loss of verticality has taken place—a futuristic cult of progress becomes inevitable. As Huston Smith has well said:

> The consequence for hope was obvious: if being has no upper stories, hope has no vertical prospect. If it is to go anywhere—and hope by definition implies a going of some sort—henceforth that "where" could only be forward or horizontal. The extent to which the modern doctrine of progress is the child, not of evidence as it would like to believe, but of hope's èlan—the fact that being indispensable it *does* spring eternal in the human breast and, in the modern world view, has no direction to flow save forward—is among the under-noted facts of history.[2]

By the end of the eighteenth century the so-called Christian humanism of an Erasmus had given way to the secular humanism of Jean Jacques Rousseau. It was an unavoidable transition: the Renaissance alliance between neopaganism and Christianity had been at best

---

2. *Forgotten Truth* (NY: Harper & Row, 1977), p. 120.

precarious and was bound to disintegrate. In retrospect one can see that the sixteenth century was in fact an age of contradiction: that the presumed union of Hellenism and Christianity had been little more than a fake. One needs but to look at the art of the period—an art which for all its genius no longer knew how to distinguish between Aphrodite and the Virgin, or between Apollo and Christ.

Yet the rise of secular humanism—with its truncated *Weltanschauung*, its "liberated" ideologies and naïve cult of progress—constitutes but half of the picture. There is another side to the story, for it happens that this transformation of Western culture was complemented by a development of the utmost importance: the rise of modern science namely. It is not surprising, moreover, that the advent of secular humanism should be accompanied by the birth of science in the contemporary sense. After all, what Huston Smith says of "hope" must apply to the sphere of knowledge as well: "if being has no upper stories," then knowledge too has nowhere to go "save forward." With the effective disappearance of verticality science becomes horizontal and turns increasingly positivistic and "operational." In other words, where man has been reduced to an earthbound creature, science in the contemporary sense becomes the only authentic and feasible way of knowing: no wonder there has been a veritable explosion in that domain!

The birth of secular science was however delayed by almost three centuries; as Nasr explains: "The scientific revolution itself came not in the Renaissance but during the seventeenth century when the cosmos had already become secularized, religion weakened through long, inner conflicts, metaphysics and gnosis in the real sense nearly forgotten, and the meaning of symbols neglected...."[3] The Renaissance had been, after all, a "rebirth"; and whereas the superior modes of Greek culture were apparently excluded from this revival, men had not yet become fully emancipated from the past. Not until the higher possibilities of the human intellect had virtually faded into oblivion could science as we know it get off the ground: not until that point in human evolution had been reached could men of intelligence devote their lives to an intellectual enterprise fully

---

3. Op. cit., p. 68.

confined to the corporeal plane of existence. The instinct in man to "look up" did not die quickly!

As Nasr notes, science in the contemporary sense began when men had quite forgotten that the things of Nature bear reference to a higher plane: to the spiritual order, no less, from whence they derive their essential content. Not only, however, did they forget, but they denied. To be precise, an anti-metaphysical bias began to emerge within the educated strata of Western society, which to this day serves as a main driving force of the scientific enterprise. It in fact defines what the man of science and indeed the man of our time perceives to be real.

Meanwhile the world—the *real* world, that is to say, the Creation — remains what it ever is. Thus, for the "pure in heart," it is still true, now as before, that "the invisible things of Him from the creation of the world are clearly seen, being understood by the things that are made,"[4] and even in this postmodernist age it is yet a fact that "the whole of the spiritual world appears mystically represented in symbolic forms in every part of the sensible world for those who are able to see" as St. Maximus the Confessor declares.[5] Truth itself does not change, nor is affected by what men believe. What has changed is man himself, or more precisely, the *psychikos anthropos*, who has all but lost his sight. To the eye of post-medieval man the natural world appears as a closed and self-sufficient realm: no longer does it point beyond itself to a transcendent domain wherein the causes of all things reside. It may of course be stretching a point to claim that "before the separation of science and the acceptance of it as the sole valid way of apprehending Nature, the vision of God in Nature seems to have been the normal way of viewing the world, nor could it have been marked as an exceptional experience" as Sherrwood Taylor claims;[6] but normal or exceptional, the experience was in any case familiar enough in medieval times. It was after all an age in which St. Hildegard could still perceive Nature as the living garment of the Holy Ghost, and could record words addressed to her by the Spirit Himself.

---

4. Romans 1:20.
5. *Mystagogy*, PG 91: 669c.
6. Quoted by S.H. Nasr, op. cit., p. 41.

For a moment, at least, let us listen to these wondrous words, which cast such an unaccustomed light upon our world:

> I am that supreme and fiery force that sends forth all the sparks of life. Death has no part in me, yet do I allot it, wherefore I am girt about with wisdom as with wings. I am that living and fiery essence of the divine substance that flows in the beauty of the fields. I shine in the water, I burn in the sun and the moon and the stars. Mine is that mysterious force of the invisible wind; I sustain the breath of all living. I breathe in the verdure, and in the flowers, and when the waters flow like living things, it is I. I found those columns that support the whole earth.... I am the force that lies hid in the winds, from me they take their source, and as a man may move because he breathes, so doth a fire burn but by my blast. All these live because I am in them and am of their life. I am wisdom. Mine is the blast of the thundered word by which all things were made. I permeate all things that they may not die. I am life.[7]

Of course, to the scientific mentality of a later age such testimony smacks of superstition, or at best is seen as "poetry" in the horizontal sense we have come to attach to that term. In the age euphemistically termed the Enlightenment, the Spirit was officially banished from this visible world, and at the hands of Galileo and Descartes even the so-called qualities have been relegated to a subjective limbo which has haunted philosophers ever since. In place of "that living and fiery essence" of which St. Hildegard had sung, the Newtonians perceived only the effects of a self-moving machine, governed by mechanical laws.

Today one knows that the Newtonian picture was but an empty dream. But though contemporary science has become incomparably more sophisticated, the universe as it appears through its lenses is no less disenchanted and de-spiritualized than it was in Newton's day: there has been no change of direction, no second thoughts in that regard. One knows today from rigorous considerations of a philosophical kind that this disenchantment of the universe is implied by

7. Quoted by S.H. Nasr, op. cit., p. 102.

the very methodology of the scientific enterprise: one can say with certainty that if there *were* "spirits" or entities of a nonmaterial kind, their existence would not register on our scientific instruments, nor show up on our mathematical maps. But this epistemological insight carries little weight: it is the thrust of modern education, after all, to convince layman and specialist alike that science *per* se constitutes indeed the one and only valid means of apprehending Nature, and that whatsoever does not appear on its prestigious charts belongs *ipso facto* to the subjective realms of fantasy.

Over a period of several centuries, and by successive stages, our civilization has come to embrace an exclusively horizontal and reductionist *Weltanschauung*. That outlook has now become the norm, the official standard in the educated world. But although the rise of an arid rationalism and a concomitant diminution of the symbolist spirit beginning in the fourteenth century may have been the primary causes that initiated this development as we have claimed, one needs also to realize that the process feeds upon itself. No sooner had the new science come to birth than it began to react upon the intellectual climate of the so-called civilized world. Before long science—or better said, scientism[8]—became in fact the dominant intellectual voice in Western civilization. It is hardly an exaggeration to say that science *cum* scientism—science with a capital S—has been the leading purveyor of "informed belief" and the driving force of cultural change since the Enlightenment. Not only, thus, has Science exacerbated the secularizing and desacralizing tendencies to which it owes its birth, but it has become itself the prime agent of that modernizing trend.

There is a conflict, an innate antagonism, then, between science and religion. Let there be no mistake about it: science is not quite as

---

8. The distinction between science, properly so called, and scientism, though generally overlooked, proves to be crucial. Briefly stated, science deals with verifiable facts, whereas scientism is a *Weltanschauung*, a kind of mythology pretending to be factual. Our dilemma is that we constantly and habitually confound the two. I have dealt with this question repeatedly, beginning with my first book, *Cosmos and Transcendence* (Tacoma, WA: Sophia Perennis/Angelico Press, 2008).

innocuous or "neutral" as many would have us believe; much less is it a spiritualizing influence. The much-disputed conflict between science and religion does indeed exist; only it lies deeper than one is apt to suppose.

The conflict—let it be said at once—is not between "scientific fact" and traditional Christian belief. We must understand, in the first place, that scientific fact, pure and simple, is virtually an unknown entity. It is hard for the epistemologically innocent to realize how much of human subjectivity has, of necessity, been incorporated into our scientific knowledge: the post-modernists have a point when they say that "facts are theory-laden." We must not forget that *the world is for us*: it is after all something that confronts us, something that we perceive and imagine, or conceive of and speculate upon. This is not to say that our knowledge is purely subjective; we are not suggesting that the world as we know it is no more than a mental construct or an apparition. What we are saying, rather, is that subjectivity and objectivity do not exclude each other: on the contrary, they go hand in hand. Though we know the world by way of mental representations, yet what we know is nonetheless the world. However, we know it not simply as it is. Inasmuch as our knowledge comes to us through an instrument (beginning with the mind), it is perforce conditioned—and diminished, as one might say—through the intrusion of subjective elements. What St. Paul has intimated with reference to our knowledge of spiritual realities applies to mundane knowing as well: "For now we see as through a glass, darkly."[9]

Science, then, is a kind of "glass." It is a glass, first of all, which filters out a major portion of the spectrum: all the higher octaves of being, one might say. And this is why *homo scientificus* cannot perceive "that living and fiery essence," and why even the familiar qualities which we do perceive with our senses are missing from what science is pleased to call "the physical universe."[10]

But the glass also functions as a lens: it magnifies and frames what it transmits. It "magnifies" in that—like a microscope—it

---

9. 1 Corinthians 13:12.
10. I have discussed this question at considerable length in Chapter 1 of *Cosmos and Transcendence*, op. cit.

brings us in touch with objective domains not otherwise known; and it "frames" in that it imposes a perspective in terms of which the given content can be received. And that is of course where subjective elements come into play.

Now this in itself does not impair the objectivity of our knowledge, as we have already said, nor is it *per se* illegitimate or harmful. If this partly subjective knowledge could be received for what it is and integrated into a respectable world-view, it would be not only innocuous and "neutral" but enlightening and perhaps even spiritually beneficial as well. As things stand, however, this is not possible. In the absence of authentic metaphysical knowledge it is virtually inevitable that the world-pictures arising from the scientific enterprise should be absolutized and consequently mistaken for the reality itself. Unlike the biologist who looks through his microscope and realizes that the ameba he perceives is *not* actually enclosed in a narrow circle, we lack the intelligence to compensate effectively for the various and sundry phantasms which the complex *modus operandi* of science has superimposed upon the universe. To sort these things out is more, surely, than most of us—including the scientist himself!—can handle. And let us not forget that the scientific development is itself driven by anti-metaphysical tendencies: how then can one expect prodigies of metaphysical discernment, be it from the scientific community or from a populace indoctrinated from infancy with scientistic beliefs? Is it surprising that the artful and fragmentary "models" which enshrine our scientific knowledge—or better said, their popularizations—should be regarded by most as a faithful picture of the real? Thus we find ourselves enclosed in a drab and meaningless universe, a world consisting of empty space and mysteriously blurred particles whirling perpetually to no discernible end, a universe in which the great truths of religion cannot but appear strange and suspect in the extreme.

"Science is our religion," Theodore Roszak has wisely said, "because we cannot, most of us, with any living conviction *see around it.*"[11] The scientific community itself does all it can to prevent us from ever doing so. By means of an education that predisposes

---

11. *Where the Wasteland Ends* (Garden City, NY: Doubleday, 1973), p. 124.

against all traditional beliefs, through the relentless popularization and vulgarization of its own theories, through the cumulative impact of technology and the artificial environment, and finally through the systematic and well-subsidized channeling of our collective intellectual resources, it has established itself within our civilization as the prime authority in high matters and sole purveyor of knowledge concerning the nature of all things.

But even that is not all. For as Roszak likewise points out: "Soon enough the style that began with the natural scientist is taken up by imitators throughout the culture."[12] Before long that prestigious "style" imposes itself upon virtually every sphere of human endeavor; and so, by stages and degrees, the scope of the scientific enterprise expands so as to encompass the full gamut of human activity, from science in the authentic sense all the way to the banal and the absurd. Sooner or later every aspect of culture is affected, every traditional norm abolished, every "pre-scientific" discipline corrupted. The higher—the more sacred!—the enterprise, the more it is debased, vulgarized, and turned upside down. It is also surely one of the "undernoted facts of history" that science inevitably begets pseudoscience and its own brand of sorcery.

We need now to take note of the fact that contemporary civilization, in keeping with its humanist and scientistic pretensions, has evolved a thoroughly skewed perception of history. Science and humanism—the twin pillars of modern culture—have joined forces to discredit and cast aspersions upon our pre-modern past. In essence we have been told that whatsoever cannot be caught in our contemporary scientific nets does not exist, and what does not conform to our humanist criteria of worth is *ipso facto* bereft of value.

An exception in our evaluation of the past is generally made when it comes to the artistic and literary spheres, or what we take to be

---

12. Op. cit., p. 31.

high insights of an ethical kind; but in matters of objective truth and "social values," it is generally taken for granted that our ancestors were "unadvanced" and rather wide of the mark. They were children and dreamers, one thinks, because they believed in realities which do not show on our scientific charts, and "unliberated" because they chose to honor the mandates of a supra-human God. "One need only ponder," writes Roszak, "what people mean in our time when they counsel us to 'be realistic'. They mean, at every point, to forgo the claims of transcendence, to spurn the magic of imaginative wonder, to regard the world as *nothing but* what the hard facts and quantitative abstractions of scientific objectivity make it out to be."[13]

It is hardly surprising, therefore, that looking back into the more distant past, the sophisticated individual of our day should behold mainly ineptitude and superstition. As Maximilian Hasak once put it: "The greater the ignorance of modern times, the deeper grows the darkness of the Middle Ages."[14]

From a less provincial point of view, on the other hand, one perceives that it is *we* who live in considerable darkness. Our collective intellectual horizon has in reality become greatly constricted, and the contemporary plethora of "hard facts" and quantitative abstractions cannot compensate us for this loss. On the contrary: as Frithjof Schuon points out, such knowledge does not enrich but rather impoverishes.[15] And Seyyed Hossein Nasr is certainly not exaggerating when he writes:

> There seems to be in this movement from the contemplative to the passionate, from the symbolist to the factual mentality, a fall in the spiritual sense corresponding to the original fall of man.... He has lost a paradise of a symbolic world of meaning to discover an earth of facts which he is able to observe and manipulate at will. But in this new role of a "deity upon earth" who no longer reflects his transcendent archetype he is in dire danger of being devoured by this very earth over which he

---

13. Op. cit., p. 124.
14. Quoted by Ananda Coomaraswamy in *Christian and Oriental Philosophy of Art* (NY: Dover, 1956), p. 29.
15. *Light on Ancient Worlds*, op. cit., p. 44.

seems to wield complete dominion, unless he is able to regain a vision of that paradise he has lost.[16]

Meanwhile our humanist and scientistic gurus continue to proclaim the gospel of Progress. The idea of future progress has moreover been joined to a systematic devaluation of our pre-modern past, which is to say that the contemporary apostles of progress busy themselves not only with the building up of a new order, but equally with the destruction of whatever still remains of the old. The ideal of progress has thus turned subversive: revolutionary in fact.

We will not speculate as to what might be the ultimate source of that boundless fascination—this veritable mania, one is tempted to say—which has apparently taken hold of the activists, the dedicated leaders in this worldwide movement. Suffice it to say that the rank and file is still somewhat passive and lukewarm, and thus needs constantly to be prodded by progressivist propaganda of one sort or another.[17] It appears that there are two main "arguments" which in the eyes of the public at large lend credence to the dogma of progress: the first are the miracles of technology; and the second is the Darwinist theory of evolution, perceived as a scientific truth.

The technological argument is hard to refute, and few there are, one fears, who remain unpersuaded. Even men of the cloth well versed in Holy Writ, including eminent theologians, have apparently been disarmed and carried off by that argument. Could it be that the feats of technology which surround us these days may indeed be "signs and wonders, to seduce, if it were possible, even the elect"? The notion is by no means far-fetched; for it cannot be denied that technology has become the number-one producer of "signs and wonders" in our contemporary world, and that "the technological repertory of the artificial environment takes the place of the miraculous"[18] as Roszak points out. The prodigies of our science-based technology seem to be indeed the only miracles in which we still believe with any real confidence, and it is the modern super-state that pulls them off.

16. Op. cit., pp. 37–38.
17. See for instance Jacques Ellul, *The Technological Society* (NY: Knopf, 1964).
18. Op. cit., p. 125.

The second major argument in support of what might almost be termed "the cult of progress" is Darwinism, as we have said, decked out as a scientific discovery resting upon "incontrovertible evidence." After all, who can ever argue with scientific facts, let alone when no one has the least idea what they are! Such a notion of science-based infallibility, once drilled into the collective psyche—from grade school up—is hard to dislodge. And if indeed we have evolved from primate stock, then on even the most pessimistic evaluation of our present status there has doubtless been progress. It is now but a small step to the conclusion that our more distant forebears—especially those with whose views we disagree—had not yet altogether cast off their simian vestiges. And so we arrive at last, by seemingly scientific and sober considerations, at the credo of Progress in its full-blown format.

Almost! It was left to Teilhard de Chardin, the scientist-theologian, to take us the rest of the way: from that secularist credo of Progress to Point Omega, the New Eschaton. Teilhard takes over—lock, stock and barrel—all the humanist and scientific conceptions of our day, and adds a few but decisive touches of his own. His great gift is to synthesize and epitomize the aforesaid conceptions, and bring the essential to a sharp point. Yet he does not only synthesize: he magnifies and exalts; in fact, *he deifies*.

His enthusiasm for science, first of all, is unbounded. Most men admire science; but Teilhard goes into ecstasy. Most believe that science is a good thing, or perhaps even an incomparable boon to mankind; but for Teilhard it is "the source of Life."[19] Scientists refer occasionally to the joy of scientific discovery; but Teilhard exults in "the divine taste of its fruit"[20]—a phrase, incidentally, which from a Biblical point of view is not without interest.

To anyone who has read his way through the Teilhardian corpus it must be clear that this pervasive predilection for unrestrained

---

19. FM, p. 20.
20. HE, p. 165.

eulogy when it comes to science is not just a literary mannerism—not simply what Medawar calls "that tipsy, euphoric prose-poetry." We most certainly need to take Teilhard at his word when he tells us that "research has for long been considered by man an accessory, an eccentricity or a danger. The moment is approaching when we shall perceive that it is the highest of human functions."[21] He is absolutely serious. Scientific research is not just sublime, or useful, or wonderful—to him it is indeed "the highest of human functions." And let us not fail to observe what this implies: as the highest of human function, scientific research is the religious function *per se*. Science, then, in its full-blown format, is the true and ultimate religion, the quest which replaces and fulfills all that men have previously designated by that term. Teilhard himself, moreover, confirms this startling conclusion when he speaks of the role science is destined to play: "It will absorb the spirit of war and shine with the light of religions."[22] What Teilhard is saying, in plain words, is that Science will eventually become the world religion, which incorporates into itself all that was true in the religions of the past. He goes so far even as to clothe this prophecy in Christian garb: "But let there be no mistake," he declares. "He who wishes to share in this spirit must die and be reborn...."

We shall deal with the question whether "this is still, of course, Christianity" at length in the final chapter. Meanwhile a simple observation might not be out of place: there is nothing in the teachings of Christ to suggest, however remotely, that the way of salvation might have something to do with natural philosophy or "scientific research." On the contrary, Christ seems to teach just the opposite: "Verily I say unto you: Whosoever shall not receive the kingdom of God as a little child shall not enter therein."[23] Think of it: to receive the kingdom of God "as a little child"—the expression would hardly apply to a modern research team! To put it as impartially as possible: the Biblical and the Teilhardian teachings do not concur on this point. What, then, are we to do? Should we cast our

21. Ibid., p. 38.
22. Ibid., p. 18.
23. Mark 10:15.

180　THEISTIC EVOLUTION: THE TEILHARDIAN HERESY

lot with Teilhard's "religion of science" or stand with St. Paul when he declares: "Where *is* the wise? where *is* the scribe? where *is* the disputer of this world? hath not God made foolish the wisdom of this world?"

Teilhard's view of science dovetails with his view of history—and, of course, Evolution. In his eyes science has existed from the start in a kind of embryonic form. It supposedly constitutes the prime vocation of man, the enterprise which beyond the simian stage defines the tangent vector, if you will, of the evolutive trajectory: "As soon as man was man" we are told, "the tree of science began to grow green in the garden of the earth. But only slowly and much later did it flower."[24] In point of fact, of course, the tree of science—as Teilhard understands the term—did not begin to flower until Sir Isaac Newton appeared upon the scene, which seems to account for Teilhard's distinctly low estimate of pre-modern civilization. Where mathematical physics—along perhaps with paleontology—has become the gauge of enlightenment, our pre-Newtonian ancestors do not fare well. And Teilhard leaves little doubt on that score:

> Yet this must be said, to our own honor and that of those who have toiled to make us what we are: that between the behavior of men in the first century AD and our own, the difference is as great, or greater, than that between the behavior of a fifteen-year-old boy and a man of forty. Why is this so? Because, owing to the progress of science and of thought, our actions today, whether for good or ill, proceed from an incomparably higher point of departure than those of the men who paved the way for us towards enlightenment.[25]

It is noteworthy that Teilhard has singled out the first century AD. in this comparison: the age when Christ walked upon the earth and His disciples recorded what has become the New Testament. If

---

24. HE, p. 165.
25. Ibid., p. 18.

the Apostles—and one fears, Christ too!—may be compared to a fifteen-year-old boy, is it any wonder that "the Old Church" stands in need of "new foundations"?[26]

But there is more. Not only was the man of bygone times inferior to us, but whatever of value he did possess is automatically passed on to later generations. There is supposedly a mechanism of "social inheritance" which ensures that the highest achievements of one generation are faithfully transmitted to the next: "Plato and Augustine are still expressing, through me, the whole extent of their personalities"[27] Teilhard tells us, quite seriously!

Now the one thing that can be said in defense of this obviously far-fetched claim is that such a mechanism of transmission does actually exist in the scientific domain: it belongs in fact to the very conception of science in the Baconian sense we have made our own. Scientific knowledge, as Sir Francis conceived of it, is something that is shared and transmitted. It is a cumulative and public kind of knowledge, gleaned through the labors of innumerable individuals and made available in libraries and computer banks. One senses that it is also an inherently quantitative kind of knowing—in its content as well as in the manner of its acquisition. "And too, as Bacon predicted," writes Roszak, "we arrive at a vast and proliferating research—a 'knowledge explosion'—under the auspices, for the most part, of just such small and ordinary minds as he foretold us would be capable of utilizing his method."[28]

Incidentally, it may be worth pointing out that Bacon ranks high among the intellectual forebears of Teilhard de Chardin, and that there is more than a passing resemblance between the two. According to Charles Gillespie, Bacon was the prophet, not so much of science as such, but of "an image of scientific progress which has been vastly more popular than science itself can ever think to be."[29] And let us not forget that in the *New Atlantis*, Bacon's third and last *magnum opus*, his vision of the scientific utopia begins to assume

26. FM, p. 23.
27. Ibid., p. 18.
28. Op. cit., p. 157.
29. Quoted by Roszak, op. cit., p. 137.

apocalyptic dimensions; as Roszak observes: "Bacon was among the first Europeans to identify the secular future as the New Jerusalem."[30]

But getting back to Teilhard's "social inheritance": the idea does in fact apply to the contemporary scientific enterprise. And if it be supposed that this is all that counts, then it *is* true that the highest attainments of mankind are being passed on and added to constantly. But even this would not mitigate the plain silliness of Teilhard's remark regarding "Plato and Augustine": for it is hardly necessary to point out that the accomplishments of these two men—let alone "the whole extent of their personalities"—have little to do with the Baconian quest, to say the least.

What Teilhard has evidently failed to grasp is that the modern scientific enterprise is not simply the culmination of a development that began "as soon as man was man," but constitutes a new start and a radical departure from the perennial ways of knowing. Admittedly, there is a certain continuity: Kepler, for example, who was quite a Pythagorean, discovered laws of planetary motion still found in physics and calculus textbooks, and as is often pointed out, our chemistry has evolved out of alchemy. Yet something essential has been lost; and in a way everyone acknowledges this fact: one knows very well that our modern sciences are nothing like the old. But needless to say, in keeping with the prevailing Zeitgeist we generally take it for granted that whatever may have been discarded or forgotten was fundamentally worthless anyway: a mere superstition as one likes to say. And this assessment is almost correct: a superstition, after all, is "any belief or attitude that is inconsistent with the known laws of science *or with what is generally considered in the particular society as true and rational*" (Webster). Only when it comes to what might legitimately be termed the traditional sciences there is in reality no actual conflict with the "known laws" of *our* science: even Ptolemy has never been disproved![31] But whatever the *modus*

---

30. Op. cit., p. 137.
31. I have dealt with this issue in "The Status of Geocentrism," *The Wisdom of Ancient Cosmology* (Oakton, VA: Foundation for Traditional Studies, 2003). See also my response to Stephen Hawking in *Science and Myth* (Tacoma, WA: Sophia Perennis/Angelico Press, 2012).

operandi of the ancient sciences may be, they obviously were not concerned with operationally definable quantities and positivistic laws.

We need also to realize that the traditional sciences were an integral part of a culture that could be characterized as religious, spiritual and metaphysical in its primary orientation. Astonishing as it may seem in this day and age, there was an intimate connection between traditional science and sacred art—all art, in fact, including architecture and music—and as late as the fourteenth century Jean Mignot, the builder of the Milan cathedral, could say: "*Ars sine scientia nihil*" ("Art without science is nothing"). Few things, perhaps, seem more incongruous to the modern mind than the notion that beauty should have something to do with truth: with *scientific* truth, no less! It appears that the very conception of "culture" in the traditional sense has disappeared from the intellectual horizon of our age—what to speak then of its reality.[32]

The fact is that much been lost, to say the least: the mechanisms of "social heredity" do not function quite as well as Teilhard would have us believe. What holds true of our contemporary sciences does not apply to all culture and all forms of human knowing across the board, and it is hardly needful to point out that Baconian science has its boundaries, invisible as these may be to the scientific mind. And however sublime, marvelous or technologically useful that Baconian enterprise may be, Teilhard's contention that it sums up and incorporates within itself all that is highest in the cultural and intellectual history of mankind is patently absurd.

What impedes Teilhard's perception of history is not just the typical mind-set of our age: Teilhard has outstripped his contemporaries in that regard. He has in fact applied himself to the task of narrowing our purview by scientific means, and has fashioned a theoretical

---

32. On the subject of traditional culture we would refer the interested reader especially to the pertinent works of Ananda Coomaraswamy, Titus Burckhardt, Martin Lings, and Seyyed Hossein Nasr.

instrument to accomplish that reduction with the utmost precision. Now, as one can readily surmise, that marvelous instrument proves to be none other than his celebrated "Law of Complexity." It is by means of this imagined Law that Teilhard has in effect cut down our field of vision to dimensions of smallness never before conceived: to a single one-dimensional continuum, in fact, coordinatized by a postulated "parameter of complexity."

Think of it: all that exists, be it in the external universe or in the realm of consciousness, is to be measured by this single stipulated parameter. And not just "measured" in the ordinary scientific sense, which is after all *quantitative*: Teilhard's parameter, supposedly, not only measures quantities, but qualities as well! The worth of an organism or of a civilization, thus, is allegedly given by its complexity. It is safe to say that no one in the history of the world had ever even remotely conceived of such a thing. The notion is as mind-boggling as it is, finally, vacuous: for as we have seen in Chapter 3, that literally marvelous parameter does not in fact exist.

Yet Teilhard is bound and determined to perceive the entire pageant of history within the preconceived framework of his stipulated Law. A single parameter of so-called "complexity" is supposed to measure the height and depth of human culture—inclusive, even, of the religious and spiritual spheres. The fundamental idea is simple enough: the universe begins with scattered particles. Then gradually, and mostly by chance, these particles come together to form aggregates. First come atoms, then simple molecules, then bigger and more complex molecules; then come cells and simple multi-cellular organisms. Moreover, in parallel with this progressive complexification, there is a concomitant rise of consciousness. Somehow complexity begets consciousness: that is the idea. And finally we come to man—presumably the most complex and most highly conscious of creatures thus far evolved—at which point evolution seems to have ceased. At least most Darwinists have thought so, and even Teilhard admits that there is not the slightest reason to believe that *homo sapiens* have become biologically upgraded since the species first appeared. Now, it is precisely at this juncture that Teilhard introduces a second idea crucial to his theory, which in a way is complementary to the notion of his Law: the idea, namely, that the formation of human collectivities—what he terms socialization—constitutes actually a *biological* process which continues the evolutive ascent.

Having first of all persuaded himself that his imagined Law has been somehow verified—by means of the same "mountains of evidence," presumably, which are supposed to have established organic evolution as a "scientific fact"—he now claims that this Law guarantees a further evolution: not of *homo sapiens* as such, but of the human collectivity. Henceforth it is supposedly the process of socialization that is building up ever more complicated aggregates, and will in course of time give birth—if all goes well—to a collective super-organism endowed somehow with a supra-human consciousness.

We will examine this startling theory with due care in the following chapter; but for the moment a few observations will suffice. What mainly complexifies the world, according to Teilhard de Chardin, is the progress of science and technology. He is thinking in terms of such things as research institutes with their burgeoning libraries, of giant interconnected industries, of telephone lines and radio-waves transmitting tons of information around the globe. But as we have noted before, Teilhard is looking at only one aspect of human culture: the most external and ontologically inessential, no less. To fit all of human culture into the ambit of his postulated Law, he must first cut that culture down to size by the gratuitous assumption that it does in fact fit: in a word, he proves that culture reduces to complexity by defining it in terms of that very complexity.

What this means is that Teilhard has systematically excluded from his purview all qualities, all art, all symbolism, all vertical reference of any kind. Now admittedly this in itself is not *per se* illegitimate: it is, after all, what a scientist does *qua* scientist. What on the other hand renders Teilhard's position not only inadmissible but indeed absurd is the contention that the resultant picture brings into view what is essential in the human phenomenon, and exhibits the true mechanism behind the dynamics of world history.

Let us not fail to observe, however, that unfounded and incongruous though it be, this Teilhardian contention has obviously been persuasive. And the reason is not far to seek: by seizing upon the notion of complexity as if it were the single all-important factor, and pretending that this concept is capable of being quantified, Teilhard has laid the foundations of a pseudo-doctrine which purports to provide a *scientific* basis for our humanist dreams. And not just our humanist dreams, but even our loftiest ideals and spiritual aspirations have been supposedly justified on this new footing: after millennia of groping, at last we *know*!

This is the great Teilhardian promise; and it is reputedly secured by his Law of Complexity. This presumed Law is supposed to hold the key to the problem of life. On the strength of this ostensibly scientific principle Teilhard would have us believe that our civilization can complexify itself, not just into a humanist utopia, but right into the New Jerusalem.

We need not detain ourselves too long with the observation that the actual facts point very much in the opposite direction. Teilhard himself, moreover, admits as much when he writes, in one of his letters,[33] that "I feel resolved to declare myself a 'believer' in the future of the world, *despite appearances*"[31]; or when he observes that "at close quarters and on the individual level we see the ugliness, vulgarity and servitude with which the growth of industrialism has undeniably sullied the poetry of primeval pastures."[34] Yet elsewhere, when the Muse of Progress beckons, he seems literally to gloat over the fact that "mechanized masses of men have invaded the southern seas, and up-to-date airfields have been permanently installed on what were until yesterday the poetically lost islands of Polynesia."[35]

Teilhard seems to forget, moreover, that what is being systematically obliterated is not just "the poetry of primeval pastures," but every last vestige of pre-modern culture. The sad fact is that "mechanized masses of men" do not simply build airfields: they bulldoze entire civilizations. And whatever might be the gain, it cannot be denied that the presumed advantages are compensated by a cultural impoverishment and an irreparable loss. It is not just idle curiosity that drives modern city-dwellers to jet across the seas

---

33. *Lettres de Voyage*, op. cit., p. 107. The italics are mine.
34. FM, p. 261.
35. Ibid., p. 131.

to spend a few days on whatever yet remains of "the poetically lost islands of Polynesia," or in some other still "unspoiled" region of the globe. Teilhard conveniently forgets that man does not live by such things as "airfields" alone. And needless to say, he forgets too that industrial civilization is coming more and more to resemble an avalanche in process of breaking loose, and that from a great many points of view that marvelous process of ever-increasing complexification threatens not just the well-being of mankind, but its very survival.

It cannot be said, moreover, in any absolute sense, that the "social organism" is becoming complexified. In the scientific and technological spheres, of course, an ongoing and progressive complexification is very much in evidence; but in other domains of culture the very opposite is taking place. In conformity with our egalitarian leanings, for example, the ancient forms of hierarchic order are in process of being dismantled, and it appears that the notion of a "classless society" has established itself just about everywhere as the ultimate desideratum. In recent decades, moreover, even the division of the sexes—which in fact mirrors the primordial duality itself—has become a prime target of egalitarian zeal, and the charming prospect of "unisex" is now before us. Despite an increasing vocational specialization, therefore, associated with the technological advance, it is plain to see that human society has become levelled-out and homogenized in other respects. And here again one may conclude that this transition is inevitable: as our *Weltanschauung* flattens, so does our culture. More precisely, civilization flattens by losing its verticality, its true hierarchic order and qualitative differentiations; and needless to say, that loss is in no wise compensated by a technology-based complexification, which truly does pertain to the quantitative domain. The world flattens, and as it flattens, it expands. This is essentially the point Huston Smith has made: when being loses its "upper stories," human culture has nowhere to go but "forwards." Under such auspices, science and technological conquest remain basically as the only viable frontier.

⊕

But this entire state of affairs is unnatural, and dangerous in the extreme. One can see—in the light of what has been said in Chapter 7 concerning the ongoing Fall—that the erosion of the traditional social order and all religious ties threatens to cut mankind off, more completely than ever before, from the primordial Center and Source of life. Having all but severed our spiritual moorings, we stand helpless in the face of that ominous "attraction of the periphery" against which there is no human or "secular" defense. And the effects of that Force are visible everywhere: all indicators point to the fact that civilization is now set upon a self-destructive course.

But whereas the presiding experts have not a clue regarding the metaphysics that stands at issue, just about everyone is beginning to sense this fact. A pronounced disenchantment with the idea of scientific progress has begun to set in. Having but recently experienced the horrors of two world wars, and finding ourselves condemned to live henceforth under the shadow of ecological doom and nuclear holocaust, we are anxious as never before. The euphoria of the nineteenth century has clearly departed from the present generation, presumably never to return. Uncertainty and frustration bordering upon despair have begun to take hold of our post-Christian civilization, and underneath such progressivist hoopla as yet remains one senses a growing *Angst*. Our literature is full of this mounting unease, and its signs are etched on the faces of even the young.

And that is where Teilhard de Chardin enters upon the stage: against this darkening backdrop of gathering clouds, the priestly scientist declares himself a "believer" in the future of the world *"malgré les apparences"*: despite all appearances to the contrary. What is more, he claims to have validated our modernist hopes and even our spiritual aspirations by the one thing in which our disillusioned civilization still believes with unabated vigor—which is *scientific law*. Now, what could be more attractive to a post-Christian generation, or to a society beginning to sink into the quicksand of despair, than a gospel of Hope and Progress, based, not upon Revelation or faith, but upon the incontrovertible evidence of Science itself! Rendered invincible, as it were, through the possession of his magic Law, Teilhard presents himself as the appointed prophet and champion of mankind.

Never mind that this Law operates with a fictitious parameter, that it abstracts from reality and ignores what is most essential; these are technical points which seem never to have bothered the multitude of Teilhardian enthusiasts. It matters not that, scientifically speaking, all this is empty talk. If it be the style that creates "the illusion of content" as Medawar says, it is doubtless "hope's élan" that inspires belief.

9

SOCIALIZATION AND SUPER-ORGANISM

IT IS ONE OF Teilhard's most astonishing ideas that the formation of social aggregates constitutes a *biological* phenomenon, which supposedly continues the process of organic evolution. "We see Nature combining molecules and cells in the living body to construct separate individuals," he tells us, "and the same Nature, stubbornly pursuing the same course but on a higher level, combining individuals in social organisms to obtain a higher order of psychic results. The processes of chemistry and biology are continued without a break in the social sphere."[1]

Such is the theory, the claim; but where is the proof? On what grounds has Teilhard put forth this gigantic claim? And as one may have come to expect by now, it turns out, once again, that Teilhard's apodictic pronouncement is buttressed by nothing more substantial than an assortment of metaphors. He speaks, for example, of "the development, through the increasingly rapid transmission of thought, of what is in effect a nervous system, emanating from certain defined centers and covering the entire surface of the globe"[2]— as if such things as telephone wires and radio transmitters could constitute an actual nervous system simply by contributing to "an increasingly rapid transmission of thought." Admittedly there is a functional analogy: both telephone wires and neurons transmit information of some sort. Yet one must obviously not press such correspondences too far; an airplane, after all, is in a certain respect analogous to a bird; but does this mean that it also lays eggs? If there

1. FM, p. 136.
2. Ibid., p. 137.

are analogies between the artificial and the biological realms, there are also differences of the most fundamental kind; how, then, could anyone be so naïve as to regard the products of communication technology as a *bona fide* nervous system? Yet this is precisely what Teilhard does when he announces that "the processes of chemistry and biology are continued without a break in the social sphere." What he means, quite plainly, is that a radio transmitter is not just analogous in certain respects to a nervous system, but that it *is* a nervous system or component thereof.

No wonder Henri de Lubac has tried to play down this amazing contention as well, notwithstanding the fact that it happens to be crucial to the Teilhardian theory. This is not a secondary or negotiable issue: the biological interpretation of socialization and technological progress is nothing less than the key notion upon which Teilhard's entire doctrine is staked. It is the bold extrapolation, validated supposedly by the Law of Complexity, on which Teilhard would found his social, political, religious, and even mystical speculations. As he himself informs us in his major work: "All the rest of this essay will be nothing but the story of the struggle in the universe between the unified *multiple* and the unorganized *multitude*: the application throughout of the great *Law of complexity and consciousness*."[3]

Enough has already been said regarding this "great Law," especially in Chapters 3 and 4. To put it bluntly: there *is* no such Law. And to be sure, the same applies to Teilhard's amalgamation of the technological and biological domains: this extrapolation from his professed Law is spurious as well. Unwittingly or not, as the case may be, Teilhard has transported us into the realm of fantasy: in these prosaic and disillusioned times he has regaled us with a contemporary version of the Prometheus myth, to the delight of millions. Or as the French biologist Louis Bounoure observes in more prosaic terms: "The presumed Super-evolution is nothing but the result of a puerile confusion between biological evolution and human progress."[4]

3. PM, p. 61.
4. *Recherche d' Une Doctrine de la Vie* (Paris: Laffont, 1964), p. 155.

Even so it will be instructive to examine Teilhard's Super-evolution in some detail. To begin with, it behooves us to take note of another basic contention: not only does the aggregation of "human particles" give rise to a biological super-organism, but what is perhaps even more surprising, the process "personalizes" these so-called particles (whatever that might mean). "The tightening network of economic and psychic bonds in which we live and from which we suffer," Teilhard explains, "the growing compulsion to act, to produce, to think collectively, which so disquiets us—what do they become... except the first portents of the super-organism which, woven of the threads of individual men, is preparing (theory and fact are at one on this point) not to mechanize and submerge us, but to raise us, by way of increasing complexity, to a higher awareness of our own personality?"[5] In a word, "socialization personalizes": so goes the formula.

But in the first place: what exactly does this mean? It is clear from the start that in Teilhard's theory the person can be nothing other than the center (real or imagined) of reflective consciousness: it can only be the ego, the subjective pole of our thought and feelings. Teilhard is still moving, as one can see, within the magic circle of the *cogito ergo sum*. Admittedly he has no more use for the Cartesian *res cogitantes* or "thinking entities," a "static" notion which has no more place in an "evolutive" universe. In Teilhard's eyes, the ego is not so much a being as it is a process: it is essentially an act. As he tells us in *The Phenomenon of Man*: "The ego only persists by becoming ever more itself, in the measure in which it makes everything else itself. *So man becomes a person in and through personalization.*"[6] But one must remember that the ego "makes everything else itself" precisely by way of thought. And so we find that the *cogito ergo sum* is still in force, and in fact holds true in a more radical sense: for whereas, in the philosophy of Descartes, that formula is understood to mean that thought implies the existence of a

---

5. FM, p. 120.
6. PM, p. 172.

thinker, in Teilhard's philosophy it means that the thinker is nothing but the center of the thought.

We still need to ask ourselves, however, what "personalization" is supposed to mean when the term is applied to "*human* particles": to "particles," therefore, which are already personalized. In other words: what does it mean to personalize a person? Once the miracle of reflective consciousness has taken place, what more can one ask? Nothing, it seems, other than to demand that the mental outreach should widen, that the human ego-in-progress should "make everything else itself" to an ever higher degree. And this is presumably what takes place through the advance of science and technology in a society dedicated to the Baconian enterprise. This is what takes place, for instance, when we watch the evening news and see distant events unfold before our eyes. And let us not forget that these things are brought into our consciousness, to be "made ourself" in thought, by way of that "collective nervous system" of which we spoke earlier.

There is a certain logic, then, in the idea that "socialization personalizes." Within the framework of the Teilhardian theory such a phenomenon, or such an effect, can indeed be envisaged up to a point. But not to the full extent Teilhard would have us believe! Under Teilhardian auspices, for example, there can be no question of a personal immortality, a *bona fide* survival of Peter and Paul; and as we have noted before, even the survival of homo sapiens beyond the habitable lifespan of the planet is also *de facto* unthinkable.

But apart from the question of survival, the Teilhardian notion of a progressive "personalization" proves to be deceptive in itself. The concept is after all exceedingly artificial and rather trivial at that. It can hardly be denied that the cultural value of such things as global news coverage is limited at best, and even the staunchest advocate of "progress" might hesitate to contend that TV heightens "terrestrial consciousness." We surmise therefore that the Teilhardian concept of "personalization" could actually arouse little enthusiasm if it were not for the fact that the term tends to be misunderstood and overestimated. Consciously or unconsciously as the case may be, Teilhard is utilizing to his advantage certain overtones or associations which have no more place in his system. We must remember that since time immemorial mankind had entertained a vastly more profound conception of personhood, a conception laden with associations which still presumably resonate to some degree in our soul. And therefore, whether one speaks of "person" or of "personalization," the very word has yet a certain ring that inspires and exalts. The word has in fact spiritual overtones which refer to another domain entirely.

It is worth recalling in this connection that according to Christian doctrine personhood constitutes actually an *imago Dei*. The human person is thus incomparably more than the so-called ego, the "I" which is inextricably tied to our thought and feelings. One could put it this way: whereas thought depends upon the person, the authentic person does not depend upon thought. And that is of course the reason why the person as such can survive the catastrophe of death, by which, after all, the brain—the instrument of thought—is unquestionably destroyed.

We need first of all to understand that the person in us did not—and could not—come into existence as a result of an evolutive process, an "aggregation" of some sort. Our bodies, perhaps: but not our personhood. "And the Lord God formed man of the dust of the ground, and breathed into his nostrils the breath of life; and man became a living soul."[7] To say that our bodies were formed "of the dust of the ground" does in fact suggest that they came into being through a process of aggregation; but it is all the more noteworthy that this alone does not suffice to make a man, and that Adam did not in fact become "a living soul" until God had infused into him "the breath of life." The reference to "breath" entails, moreover, that what is thus infused constitutes a *spiritual* principle, and even suggests a certain continuity or kinship with the divine: what God "breathes" into Adam is unquestionably His Breath, which theologians are wont to identify with His Spirit.

But then, if our personhood did *not* originate by an evolutive process of aggregation, how can it be supposed that we shall be further "personalized" through a process of socialization conceived along evolutive lines? One can still speak, presumably, of "personal-

---

7. Genesis 2:7.

ization" in the sense of self-realization or spiritual growth, a question of bringing "image and likeness" into coincidence as the Orthodox say. But if indeed our personhood derives "from above," then it must follow that "personalization" likewise hinges upon a spiritual influence of some kind.

Under favorable circumstances, moreover, such an influence may be transmitted through the medium of a human collectivity; and such is the case in authentically traditional civilizations, and preeminently, of course, in the institutional Church. This is, after all, what "tradition" (from *tradere*, to "deliver" or "transmit") signifies. There was human failing, of course, a pervasive and recurrent falling off from the values and standards upon which traditional societies were based. And yet a spiritual and spiritualizing influence continued to flow, as it were, through multiple channels of transmission. In modern times, on the other hand, the picture has changed: what impedes us today is not simply a falling short on the part of individuals, but an inbuilt godlessness, a Promethean spirit of disobedience officially instituted and inscribed, as it were, upon our tablets of law. A kind of collective neo-humanist hubris insulates us *en masse* from the spiritual world. It is not a question of having outgrown the past, as one likes to believe, but of spiritual incomprehension and concomitant rebellion. In the emerging Brave New World the breaking of sacred vessels has come to be regarded as a meritorious act and a mark of enlightenment. My point, then, is that in such a society "personalization" in the authentic sense is an exceedingly untypical occurrence. Yet the phenomenon continues to exist. There will always presumably be individuals who have what it takes to swim against the stream; and let us not forget that numerous traditional channels, both religious and cultural, are still operative.

It follows from these (admittedly all too cursory) observations that in Christian terms "personalization" can mean nothing more nor less than authentic spiritual growth—which is not to say, of course, that our contemporary *periti* are incapable of using the term to designate the very opposite. The devaluation and debasing of linguistic coinage has always in fact been a favorite occupation of anti-traditionalists, and is doubtless one of the most effective means of "breaking sacred vessels." In any case, within the technological society personalization has become exceedingly untypical: in a sense, and to some degree, we all participate in the ongoing collective Fall. In certain respects every member of an aberrant society suffers from the misdeeds and infidelities of those who contribute to the formation of the *status quo*. Not that we are morally responsible for the misdeeds of another; what has been placed upon us is not the onus of sin, but of a certain incapacity. Yet it is not by any means an absolute incapacity, one that cannot be overcome: that is the crucial fact. As Eric Voegelin has well said: "No one is obliged to take part in the spiritual crisis of a society; on the contrary, everyone is obliged to avoid this folly and live his life in order."[8]

Which leads me to a final observation: the highest degree of personalization is realized precisely in those heroic men and women rightfully referred to as "saints." One needs but to read an account of their lives, and a few of their words perhaps, to realize how unique and powerful a personality shines through these outer vestiges. As every flower has its own inimitable fragrance, so every saint seems to have a "spiritual perfume" all his own. And let us not forget to observe also that this spiritual growth, this authentic personalization, is achieved without any of the collective means Teilhard prizes so highly: no research institutes, no scientific congresses, no stupendous technology. As a rule these spiritual giants live as simple children of the Church. Their circle of physical contact has often been limited in the extreme; and not a few of their number have lived out their years in the isolation of deserts and caves.

The conclusion that emerges from all these considerations is clear: it is *not* true that "socialization personalizes." This, too, is an illusion, a Teilhardian myth.

Having touched upon the subject of "spiritual influences," it behooves us to recall that these can be of very different kinds. Authentic Christian doctrine affirms that there exist various angelic orders, and "fallen angels" as well. Admittedly these subjects have

---

8. *Science, Politics and Gnosticism* (South Bend, IN: Gateway, 1968), pp. 22–23.

not received much attention since the Enlightenment; yet the fact remains that "There are more things in heaven and earth, Horatio, than are dreamt of in your philosophy." Yes, not only does the order of creation comprise angelic or "spiritual" beings of various kinds, endowed with powers we humans do not possess, but it happens that not all of these are beneficent. Let us understand it well: this is not a superstition come down from pre-scientific times, but a hard truth which it behooves us to reckon with. Whether Albert Einstein, say, or Stephen Hawking know it or not, there *are* malefic spirits in the world—it is not fable. We know this not only on the authority of Scripture and Tradition, but also on the basis of hard empirical facts: one does not exorcise a fable! At the end of the day we may yet realize that St. Paul was right after all when he declared: "We wrestle not against flesh and blood, but against principalities, against powers, against the rulers of the darkness of this world, against spiritual wickedness in high places."[9]

Getting back to Teilhard de Chardin, it is clear that on this fateful issue as well the French Jesuit is solidly aligned with the neo-humanist camp. This follows already from his evolutionist premises, and also from that Law of Complexity which is supposed to be the key to just about everything: for if spirit is indeed the specific effect of complexity, then spiritual nature submits to the notions of "more" or "less," but not to *qualitative* judgments. The distinction between "beneficent" and "malefic" spirits, in particular, becomes inconceivable under these auspices. Yet all the same, value judgments do enter into the theory via a postulate in which Teilhard ardently believes: that Evolution, namely, is productive of good. If spirit, then, be the product of evolution, it is *ipso facto* desirable, *ipso facto* good: that is what it comes down to. Never mind that this theory makes no sense even from a purely scientific point of view: Teilhard speaks as if every sufficiently complex molecule were a nutrient or a vitamin, and as if there were no such things as cyanide. The more "complex" the aggregate, the better and more spiritual: that is the gist.

Let us not fail to observe that in the social and political spheres

9. Ephesians 6:12.

this means that every form of totalitarianism or collectivization—no matter how brutal—can be accorded high rank, provided only that it complexifies "the human mass" to a sufficient degree. It also means that every "center of centers" can be assimilated to Omega, provided only that it attracts "human particles" and causes them to aggregate. Within the framework of Teilhard's theory it becomes impossible to conceive of an intrinsically evil society—of a mode of social aggregation, namely, which is inherently detrimental to the well-being of its members. Where evil is conceived as nothing but disorder, order of whatever kind becomes beneficent. In the sphere of human collectivities no less than in the domain of molecules, Teilhard is committed to the risky notion that every complex is a vitamin, and that cyanide does not exist.

What Christianity refers to as "discernment of spirits" has thus been ruled out. There is no more difference between angels of light and angels of darkness; and it is doubtful, in fact, that there are any angels left at all. In a Teilhardian universe there can be no such thing as "spiritual wickedness in high places." The idea that demons not only exist, but can also congregate, has become inconceivable: in Teilhard's universe there is no room for a "synagogue of Satan." [10] What is more, Teilhard has deftly done away with Satan himself: the concept of a nether pole of attraction, an anti-Omega, has also been ruled out. As we are told repeatedly in *The Divine Milieu*, all human labors—all human acts without exception—are done under the aegis of the Risen Christ. At one point Teilhard even claims that "evil spiritual powers" [sic] are His "living instruments"[11]! Think of it: all those monsters of iniquity—from Jack the Ripper to Adolf Hitler—the "living instruments" of Christ!

The paramount desideratum in Teilhard's eyes—"the one thing needful" in fact—is the building up of the technological super-state; and from an evolutionist point of view there is a certain logic in that

---

10. Revelations 2:9.
11. DM, p. 127.

position. As Teilhard points out: "To all appearance the ultimate perfection of the human *element* was achieved many thousands of years ago, which is to say that the individual instrument of thought and action may be considered to have been finalized." As regards the human individual, in other words, evolution has reached the end of the line. "But there is fortunately another dimension," he goes on to say, "in which variation is still possible, and in which we continue to evolve."[12] Now that other dimension refers of course to the social domain, the sphere of human collectivity. From an evolutionist angle of vision there is only one viable path before us, and that is to aggregate through collectivization. To act, to produce, and even to think collectively—that is "the growing compulsion" from which there is supposedly no escape.

Teilhard realizes, of course, that to most of us the prospect of compulsory collectivization is distinctly unpalatable. As yet it is mainly the activists—the professional collectivizers one could say—who seem to be fully sold on the prospect of a joyous serfdom to the emerging super-state. The rest remain disquieted and apprehensive in varying degrees. And for good reason: if facts mean anything, there is little ground for euphoria. Even Teilhard admits that "the modern world, with its prodigious growth of complexity, weighs incomparably more heavily upon the shoulders of our generation than did the ancient world upon the shoulders of our forebears."[13] And so far as "recent totalitarian experiments" are concerned, he admits that the results have been less than encouraging.

But he argues that we must not give up that easily: "In so far as these first attempts may seem to be tending dangerously towards the sub-human state of the anthill or the termitary, it is not the principle of totalization that is at fault, but the clumsy and incomplete way in which it has been applied."[14] And what *are* these "first attempts"? As it happens, Teilhard's reference to "recent totalitarian experiments" pertains to a speech delivered before the French Embassy at Peking in the year 1945, and must consequently refer to

12. FM, p. 16.
13. Ibid., p. 122.
14. Ibid., p. 123.

the exploits of Hitler and Stalin (Mao Tse Tung's "experiments" had not yet run their course at that time). It is interesting that Teilhard seems not to be displeased with "the principle of totalization" exemplified by these events, but only censures the manner in which that principle has been applied. Teilhard does not explain in what way the exploits of Hitler and Stalin were "clumsy" and "incomplete"; was it perhaps because an insufficient number of undesirables were liquidated: five or six million apiece might not have been enough? Be that as it may, what we are told in no uncertain terms is that, sooner or later, the experiment will succeed.

It *must*, because there is no other way: that is the bottom line. Repeatedly, in fact, Teilhard goes through an elaborate dialectic to prove that all other roads are barred. He does so, for instance, in an essay entitled "The Grand Option," in which he confronts us successively with three alternatives. In each case the desired option is exhibited as the second of two possibilities, the first of which is patently unacceptable. Each choice, moreover, leads on to the next, until one arrives at the final conclusion. As Teilhard informs us at the end of this tour:

> Our analysis of the different courses open to Man on the threshold of the socialization of his species comes to an end at this last fork in the road. We have encountered three successive pairs of alternatives offering four possibilities: to cease to act, by some form of suicide; to withdraw through the mystique of separation; to fulfill ourselves individually by egoistically segregating ourselves from the mass; or to plunge resolutely into the stream of the whole, in order to become a part of it.[15]

Now, the only difficulty is that in each instance the stipulated alternative turns out to be spurious: the whole exercise is a classic example of what may be termed the method of false alternatives. It is true enough that the first three options are unacceptable; but this does not by any means establish the fourth.

Consider for example the second presumed alternative:

---

15. PM, p. 49.

On the one hand there are those who see our true progress only in terms of a break, as speedy as possible, with the world.... And there are those on the other side, the believers in some ultimate value in the tangible evolution of things ... withdrawal, or evolution proceeding ever further? This is the second choice that human thought encounters in its search for a solution to the problem of action.[16]

The crucial implication, of course, is that if we reject the first of the two proposed options—the straw man—we have *ipso facto* committed ourselves to the second. But this is not actually the case; there happens to be a middle ground between the two alternatives.

An ancient inscription to be seen at Fatehpur Sikri comes to mind: "Jesus says"—so reads the inscription—"the world is a bridge: pass over it, but build no house thereon." Here we have the answer, the Christian response no less: the world is not an evil to be shunned, nor an illusion from which to liberate oneself; but neither is it destined to be our permanent home. It is rather a bridge. There are those, perchance, who would jump off and those who wish to remain; but "Jesus says: pass over." The world is there, not just to torment or to delight, but above all to teach wisdom and kindle charity; and this it does by its cruelties no less than by its tender mercies. The Christian—yes, the Christian optimist!—has no reason to suppose that the world is destined to become better in course of time: it fulfills its purpose just as it is. It is *we*—and not the world—who need to improve. And if it should happen that as a result the world becomes better as well, there is no harm in this—so long as we remember that it still is no more than "a bridge."

The Christian is not overly concerned with the world at large, nor does he have any special allegiance to a super-state. His love extends above all to God, and in second place to his brothers and sisters in Christ. And he understands that it is men and women who are to be saved—not the universe or the human race! The Teilhardian conceptions of universal convergence and collective salvation are utterly foreign to Christianity: Jesus never taught that the world is

---

16. FM, p. 36.

destined to evolve into a paradise. He gave us to understand, rather, that this world is a bridge: pass over that bridge, and build no house thereon.

There *is* a third option, then, a way of living and acting in the world which entails neither withdrawal ("as speedy as possible"), nor yet belief in progress and commitment to the formation of a collective super-organism. And this observation, simple and obvious as it is, suffices to invalidate Teilhard's argument: his entire long-winded exhortation proves to be inconclusive, and it remains to be seen whether in fact "we can do no other than plunge resolutely forward, even though something in us perish, into the melting-pot of socialization." Nor does it help his case when Teilhard adds, in ostensibly Christian accents: "Though something in us perish? But where is it written that he who loses his soul shall save it?"[17] What Teilhard neglects to take into account is the possibility that there may indeed be more than one way of "losing one's soul."

To plunge resolutely forward "into the melting-pot of socialization": that is in Teilhard's eyes the Way of Salvation which leads straight into the New Jerusalem. And this means basically that the Marxist and the Christian are fellow-travelers: they may not know it yet, but despite certain inessential differences in their respective views they are converging to one and the same goal. As Teilhard has put it:

> Take the two extremes confronting us at this moment, the Marxist and the Christian, each a convinced believer in his own particular doctrine, but each, we must suppose, fundamentally inspired with an equal faith in Man. Is it not incontestable... that despite all ideological differences they will eventually, in some manner, come together on the same summit?... Followed to their conclusion the two paths must certainly end by coming together: for in the nature of things

---

17. Ibid., p. 54.

everything that is faith must rise, and everything that rises must converge.[18]

Let us carefully examine that momentous claim, which has evidently been swallowed by a multitude of post-Conciliar religious and laymen alike.

To begin with, we must object to the idea that what the Christian believes is "his own particular doctrine." Actually, nothing could be further from the truth; did not even Christ Himself declare: "My doctrine is not my own, but his that sent me"[19]! To set the record straight: the teaching of Christianity is *not* a man-made doctrine on a par with Marxism, but a truth revealed by God, speaking to us first through the prophets, and in the fullness of time by His incarnate Son. It is a teaching, moreover, which "flesh and blood" cannot receive; for "No man can say that Jesus is the Lord, but by the Holy Ghost."[20]

Nor can it be maintained that the Christian and the Marxist are "fundamentally inspired by an equal faith in Man." What inspires the Christian, first of all, is not in fact a faith in Man: he is by no means a humanist. The Christian believes, first and foremost, in God: in the God of Revelation, "the living God of Abraham" as distinguished from "the God of the philosophers." He does, of course, uphold a certain faith in man: he believes in the dignity of man, in certain inalienable rights, and above all, in his high calling as a potential son of God. But he does *not* believe in man as an autonomous being; not for an instant does he accept the notion that the human individual or the human collectivity is at liberty to deny God or reject His mandates. The Christian consequently abhors secular humanism, regardless of whether it be conceived along individualist or collectivist lines. His faith in man, therefore, is qualified. He knows from the start that God alone is absolutely good, and that all strength and all glory derive from Him. He knows full well that his own strength is weakness, and that human wisdom is mere folly

---

18. Ibid., pp. 198–199.
19. John 7:16.
20. 1 Corinthians 12:3.

before God. The true Christian, therefore, does not put too much stock in human means, be it the strength of numbers, the power of wealth, or the efficacy of Five Year Plans. He knows that God is able to accomplish the mightiest works through the most frail and humble instruments, and understands that whatever man does simply on his own authority and in his own name will eventually fail. Cities and empires will crumble into dust, and even our boast of scientific knowledge will prove hollow in the end; like Isaac Newton he realizes that we can do no more than gather a few pebbles by the seashore while the ocean of God's truth remains untouched.

What utter foolishness, then, to maintain that the Marxist and the Christian are "fundamentally inspired with an equal faith in Man"!

But there is still more: the Christian knows that he belongs to a Kingdom which is not of this world, nor is to be accessed at the end of the collective road; nothing could be further from the truth: "Neither shall they say, Lo here! or, Lo there! for, behold, the kingdom of God is within you."[21] Unlike the Marxist utopia, the Kingdom of God exists already and is close at hand though the world sees it not and knows it not. And the reason too the Christian knows: we do not see that Kingdom because we are looking in the wrong direction. We need a change of heart, an authentic *metanoia*—not only that we feel sorry for our evil deeds, but above all a *bona fide* act of repentance, a veritable "turning round." The call of St. John the Baptist applies to this very day: "Repent ye: for the kingdom of heaven is at hand."[22]

Meanwhile it is plain to see that the neo-humanist gurus of the future paradise are paying little heed to this call, and that the world at large is heading precisely in the wrong direction. Let us be clear on this point: the Bible speaks not just of "brotherly love," but of a few other things besides—the words *"vanitas vanitatum"* ("Oh vanity of vanities"), for instance, are also to be found in the Good Book. And harsh as it may sound to humanist ears, the primary concern of the Christian is not to assist in the accomplishment of collective projects but rather to convert souls, beginning with his own. As a rule his

---

21. Luke 17:21.
22. Matthew 3:2.

words are not pleasing to aficionados of the status quo: for it is his vocation as a Christian to convict the world of its folly. He understands full well that "the one thing needful" is indeed the salvation of our immortal soul. His charity, therefore, is something quite different from humanist philanthropy, not to speak of Marxist fanaticism: we must never forget that the primary object of authentic Christian charity is to bring souls to Christ.

Teilhard, of course, does not perceive any intrinsic conflict between the following of Christ and the way of the world. He thinks that such opposition as there appears to be is due in large part to an antiquated understanding of what Christianity is about, based upon a pre-Darwinist conception of the universe. For Teilhard it is a settled conviction that Christianity can be nothing more than one particular manifestation of the human spirit, one particular expression of a single universal human drive. "Look well," he declares, "and we shall find that our Faith in God, detached as it may be, sublimates in us a rising tide of human aspiration." And he adds: "It is to this original sap that we must return if we wish to communicate with the brothers with whom we seek to be united."[23] To this the Christian is bound to reply: Look well, and you shall find that our faith in God derives from an incomparably higher source. As with Peter, so with us: "Flesh and blood has not revealed it unto you, but my father which is in heaven."[24] And so too that faith impels us in a new direction, and towards a very different end: it wars, in fact, against the "human aspirations" of the carnal man, the man who lives on the level of "flesh and blood." St. Paul speaks in principle for all Christians when he declares: "I delight in the law of God after the inward man; but I see another law in my members, warring against the law of my mind, and bringing me into captivity to the law of sin which is in my members."[25] But what else is that "law of sin which is in my members"—what else could it possibly be—than that "original sap" on which Teilhard has his eye! Let him spin his rhetoric as he will: the fact remains that the Christian and the neo-humanist

23. FM, p. 200.
24. Matthew 16:17.
25. Romans 7:22–23.

are *not* fellow-travelers. Their respective paths and destinations are worlds apart—and each of us is obliged to make his choice.

At one point Teilhard himself raises the natural question whether the neo-humanist "religion of conquest" which he extols may not in fact be Promethean. In an essay commemorating the first explosion of an atom bomb—having informed us that "thus the greatest of Man's scientific triumphs happens also to be the one in which the largest number of brains were enabled to join together in a single organism"[26]—he proceeds to contrast what he takes to be the Promethean and the Christian spirit. Teilhard perceives the former as "the spirit of autonomy and solitude; Man with his own strength and for his own sake opposing a blind and hostile Universe; the rise of consciousness concluding in an act of possession."[27] By way of contrast he envisages the Christian ethos as "the spirit of service and of giving; Man struggling like Jacob to conquer and attain a supreme center of consciousness which calls to him; the evolution of the earth ending in an act of union."[28]

It behooves us consider these characterizations, beginning (on the side of Prometheus) with "the spirit of autonomy and solitude." What precisely does Teilhard have in mind? He is thinking no doubt of the self-centered human individual, the egoist who lives for himself and fends for himself. To such a man the universe appears "blind and hostile" because, as a matter of fact, it is not specifically designed to serve his own selfish ends. And what are these ends? Wealth and power, mainly; or, as Teilhard puts it, "an act of possession."

The Christian attitude, on the other hand, is said to be "the spirit of service and of giving," and also of course, "the spirit of love." But what *kind* of love? That is ever the crucial question. Everyone, to be sure, speaks of love; and the heretics, it would seem, do so the most. What, then, is Teilhard alluding to? Is it the true Christian *agape*? Is

---

26. FM, p. 149.
27. Ibid., p. 153.
28. Ibid.

it the love Christ enjoins upon us in "the first and great commandment"? Or in the second, perhaps: "Thou shalt love thy neighbor as thyself"? And what about the idea of "ending in an act of union": union with what? What is actually the Eschaton Teilhard has placed before us: is it God or is it Man?

By now the answer to this basic question should be plainly evident: Teilhard's gaze is fixed upon Mankind, upon the emerging super-state, upon that super-organism which is supposedly being formed by way of "planetization": "the evolution of the earth ending in an act of union" is indeed what his teaching is about.

But is this Christianity? Is this the teaching of Christ? Once again Teilhard has muddied the waters with his dialectic of false alternatives. To begin with, he has misconstrued the Promethean ethos. If it be admitted that Teilhard's egoist is indeed a Promethean type—a matter which is debatable—he represents in any case a comparatively harmless variety: there is another kind, far more dangerous, which Teilhard has left out of account. Who would deny, for instance, that the fanatical Nazi who is ready at the drop of a hat to suffer death for *Reich* and *Führer* is any less Promethean than the more familiar and far more innocuous egoist? Teilhard is very much mistaken when he identities the Promethean characteristic with the egoistic bent—as if Promethean behavior were simply a question of being selfish or antisocial. He forgets that devils, too, can collaborate. What actually characterizes the Promethean spirit—let it be clearly understood—is not the pitting of an individual human "I" against the human collectivity, but rather the pitting of man, be it as an individual or as a collective entity, against God.

To glorify oneself, or to glorify some human conglomerate: is there really quite so sharp a distinction between the two as Teilhard would have us to believe? Is there no such thing as a *family* egoism, a *national* egoism, or an *ethnic* egoism? The "I" and the group, the human collectivity with which *I* identify: are these not simply the two complementary poles (like the two foci of an ellipse) around which the carnal man invariably circulates? Is this not already implied by the fact that man is inherently a *social* creature? And is it not clear, also, that even robbers act in concert with their kind, and that even murderers take care of their own?

Teilhard confuses the Promethean as such with the egoist, forgetting that the latter embodies but one aspect of the Promethean type, and then proceeds to misconstrue the Christian as such by confusing him with the complementary aspect of the Promethean, thereby turning Christianity deftly into Marxism: it thus becomes the worship of Collective Man.

Teilhard conveniently forgets, moreover, that Prometheus stole the fire from Heaven, not just for himself, but precisely for the benefit of mankind, in perfect accord with Teilhard's own ideals. The fact is that Teilhard himself stands on the Promethean side of the divide. If there be yet any doubts on that score, the following elucidations should suffice to set these doubts at rest:

> Thus considered, the act of the release of nuclear energy, overwhelming and intoxicating though it was, began to seem less tremendous. Was it not simply the first act, even a mere prelude, in a series of fantastic events which, having afforded us access to the heart of the atom, would lead us on to overthrow, one by one, the many other strongholds which science is already besieging? The vitalization of matter by the creation of super-molecules. The remodelling of the human organism by means of hormones. Control of heredity and sex by manipulation of genes and chromosomes. The readjustment and internal liberation of our souls by direct action upon springs gradually brought to light by psychoanalysis. The arousing and harnessing of the unfathomable intellectual and effective powers still latent in the human mass. . . .[29]

All this is of course Promethean to the core. To top it all, this Faustian fantasy culminates in the following frenzied exclamation: "In laying hands on the very core of matter we have disclosed to human existence a supreme purpose: the purpose of pursuing ever further, to the very end, the forces of Life. In exploding the atom we took our first bite at the fruit of the great discovery, and this was enough for a taste to enter our mouths that can never be washed

---

29. Ibid., p. 149.

away...."³⁰ A "fruit of great discovery," a fateful "bite," a taste "that can never be washed away": where have we heard all this before? And what about "the very core of matter" wherein reside "the forces of Life": is this not indeed reminiscent of the Tree of Life "in the midst of the garden"? In opposing "the light of a growing unanimity" to "the nightmare of bloody combat,"³¹ Teilhard reminds us, moreover, that by its very nature this fateful drama of discovery pertains indeed to the perennial conflict of Good versus Evil And what, finally, is the "supreme purpose" that goads men and women into "pursuing ever further, to the very end, the forces of Life"? What is it that drives the Teilhardian Prometheus to "explode the atom," and in so doing discover "another secret pointing the way to his own omnipotence"³²? The very phrase gives the secret away. What confronts us in this itself Promethean outburst is none other than the ancient temptation, the Serpent's promise which has tantalized our race: "Ye shall be like gods." Even the Biblical plural fits: this is no solitary venture.

No one need be surprised that Teilhard exhibits a special interest in the Pauline teaching concerning the Mystical Body of Christ. Teilhard's approach to Christian dogma, as we have seen, is selective: basically, he accepts those articles of Faith which can, with suitable modifications, be fitted into his evolutionist scheme, and does away in effect with the rest. Now the Mystical Body, so Teilhard believes, can be made to fit. "To this mystical super-organism," he tells us, "joined in Grace and charity, we have now added a mysterious equivalent organism from the domain of biology: the 'Noöspheric' human unity gradually achieved by the totalizing and centering effect of Reflection."³³ What Teilhard is driving at, to be sure, is that these two "super-organisms" are in truth one and the same, and

---

30. Ibid., p. 151.
31. Ibid., pp. 151–152.
32. Ibid., p. 148.
33. Ibid., p. 232.

that it behooves us to recognize this fact: "How can these two super-entities, the one 'supernatural', the other natural, fail to come together and harmonize in Christian thought?"

It is a diplomatic way of announcing the presumed identity. We must bear in mind that when this "prophecy" was made some ninety years ago, that identification could not have passed theological muster: it required the Council and its aftermath to create a climate in which such ideas could be seriously entertained. So far as Teilhard himself was concerned, on the other hand, the anticipated "coming together" of these two "super-organisms" is a *fait accomplis*: and as a matter of fact, he gives the secret away in a footnote by referring to "the collective human organism ('the mystical body')" as if the two comprised one and the same entity. It is interesting, moreover, that what has been put under quotes is "the mystical body," as if the so-called collective human organism ("from the domain of biology") were indeed the concrete reality, to which the theological phrase must ultimately refer. And is this not also why the adjective "supernatural" has been put in quotes when Teilhard refers to "these two super-entities, the one 'supernatural', the other natural"? It seems to be Teilhard's way of informing us that the Mystical Body is supernatural only from the standpoint of a pre-evolutionist theology which had not yet come to grips with the actual facts.

Let us now examine what Scripture and Tradition have to say regarding the two "super-organisms." There are Biblical references, of course, to the Mystical Body; but are there not also allusions to that other "super-organism": the one to be built through the collective enterprise of men?

There are indeed; and the most striking, no doubt—the example which immediately springs to mind and has become proverbial—is the story of Babel: "And they said, Go to, let us build a city and a tower whose top may reach unto heaven"[34]: what could be more unequivocal than that? "A tower whose top may reach unto heaven": does this not recall to our mind the Teilhardian Omega? There is even a reference in the same pericope suggestive of "planetization"

34. Genesis 11:4.

and "unanimity," the twin notions which figure so prominently in the Teilhardian scheme: "The whole earth was of one language and one speech"[35] we read. "One language": could this not be interpreted in the present context as "the universal language of science"? And "one speech": might this not refer to a single global system of communication, the very thing Teilhard conceives as the actual nervous system of the stipulated super-organism in process of formation?

These are exegetical speculations, of course; but they are by no means incongruous. It is also of interest to observe that Genesis reveals the Promethean character of this collective venture, which consists not so much in a denial of God—Prometheus himself was no atheist!—as in the fateful notion that man, by his own collective endeavor, can accomplish what by right appertains to God. Now this putting of man in the place of God—is this not indeed the hallmark of the Promethean? And what, let us ask, is the object of this presumption: what is it that Promethean man sets out to accomplish? Is it not in fact "to build us a tower whose top may reach unto heaven"? Is not the Promethean aspiration tantamount to the idea that mankind, through collective enterprise guided by reason, can "reach up to heaven": can enter the Heavenly Jerusalem? The great presumption, then, the Promethean Sin, is not simply the hope that Heaven can be attained, but the notion—the "evolutionist faith" if you will—that Heaven can be taken by dint of collective human enterprise.

And this is what, figuratively speaking, brings down "the wrath of God": the Tower always falls, is always shattered in the end. It is built upon sand, a vain thing; and as the Good Book likewise teaches, it is not Babel, but that other City—the one that is built upon a Rock—that shall prevail and be victorious, and is destined to endure *in saeculum saeculi* ("for ever and ever").

There are two Cities, then: one built by man, the other by God or by "the sons of God," as we are also told. "The Lord came down to see the city and the tower which the children of men built"[36]: such

---

35. Genesis 11:1.
36. Genesis 11:4.

is the Biblical characterization of Babel. It is the City "built by the children of men": the "earthly city" as St. Augustine observes, built not by "the sons of God," but by "that society which lived in a merely human way."[37]

There are then two Cities: the earthly and the heavenly, the natural and the supernatural. And not only are the two different, but they are actually opposed: this, too, we learn from the Book of Genesis. God Himself takes up arms, as it were, against Babel: "So the Lord scattered them abroad from thence upon the face of all the earth: and they left off to build that city."[38] The "unanimity" of which Teilhard speaks will not last. It will eventually give way to conflict: that is the Biblical prediction. And the Tower will crumble into dust long, long before Heaven—or Point Omega—is attained.

To speak of the two Cities is to allude to the inherent opposition between the way of the world and the following of Christ. And notwithstanding its current unpopularity—its extreme political incorrectness—this recognition proves crucial: the two Cities, or the two Ways, are always with us and forever at war. There can be no reconciliation between the two, no *aggiornamento* of any kind. And the decisive fact is this: each and every one of us will have to choose to which City he would belong.

Physics teaches that bodies move invariably along the path of "least action": the path of least resistance, if you like; but the Christian does just the reverse: he moves perpetually "against the stream." The way of the world, one might say, is like that of a river which descends, as all rivers do, and broadens as it runs its course; but the way of Christianity resembles rather that of a mountain path which narrows as it ascends towards the summit.

A striking description of the two Ways is to be found in the Gospel according to St. Matthew: "Enter ye in at the straight gate: for wide is the gate, and broad is the way, that leadeth to destruction,

---

37. *City of God*, 16.5.
38. Genesis 11:8.

and many there be which go in thereat: Because straight is the gate, and narrow is the way, which leadeth unto life, and few there be that find it."[39] And let us not fail to note that in the very next verse we read: "Beware of false prophets, which come to you in sheep's clothing, but inwardly they are ravening wolves"!

No wonder the Christian life commences perforce with an act of repentance, a *metanoia* or "conversion" as we have said before. On account of Original Sin we are, by nature or by instinct as it were, headed in the wrong direction. No wonder too that the humanist mainstream leads, not to the Promised Land as the apostles of Progress would have us believe, but to a fatal precipice: as Christ has repeatedly forewarned, great will be the destruction of those who shall be swept over its fearful brink. Repent therefore and change your course while there is yet time: such is the Christian imperative.

But needless to say, that is not the teaching of Teilhard de Chardin. The closest he ever comes to admitting that there is after all a difference between the Christian and the humanist ideals is when he likens the two to vectors pointing in orthogonal directions, which need henceforth to be combined. "OY and OX, the Upward and the Forward," he tells us: "two religious forces, let me repeat, now met together in the heart of every man; forces which, as we have seen, weaken and wither in separation. . . ."[40] But this, too, is untenable. In the first place, inasmuch as authentic religion is indeed concerned with the supernatural and derives "from above"—unless, of course, we assume from the outset that its claims are invalid—it is something altogether different from neo-humanism in any of its forms, which is clearly a man-made ideology. To situate these two ideals on the same plane, therefore, and speak indiscriminately of "two religious forces" is already to obfuscate the issue. So too one may question whether Christianity is destined to weaken and wither unless it joins forces with the neo-humanist enterprise: has it not waxed strong and thrived for at least a millennium and a half on its own? And if we look carefully at the subsequent decline, do we

39. Matthew 7:13–14.
40. FM, p. 278.

not find that the very opposite has actually occurred: that the more Christianity did conform itself to humanist and scientistic trends, the weaker and more lifeless it became. It could not be otherwise, for "No man can serve two masters."[41] As Christians we are called upon to love and serve God, not just halfway, but "with all thy heart, and with all thy soul, and with all thy mind": what could be more unequivocal than that?

There can be no mixing of the Christian and neo-humanist ideals, no merging of the two vectors, as Teilhard suggests. We must choose: we are free to take either path; no one compels us. Meanwhile two voices can always be heard: there is the loud, external voice of the world calling us to that broad mainstream on which "the many" seem always to be embarked. It speaks to us of Progress, the conquest of Nature and the setting up of a Kingdom upon this earth; it speaks of Evolution, of a Promethean venture: "Ye shall be as gods." But there is also a second voice, soft and gentle, a voice to be heard in stillness and in poverty of spirit: "Come unto me, ye who labor and are heavy laden" it whispers. And that still voice calls us, not to the Promethean enterprise of those who would be like gods, but to a spiritual heroism so abundantly and gloriously exemplified by the saints and martyrs of the Church.

These are the two options, the two paths. And so far from being destined to converge, the inherent opposition between the two will become only more acute as we approach that limit-point of history which Christianity knows as the Parousia, the Second Coming of Christ. "And ye shall be hated of all men for my name's sake" Christ declares with reference also to those latter days: "but he that endureth to the end shall be saved."[42] We are given to understand, in the apocalyptic discourses of Christ as well as in the Book of Revelation, that there is to be a vast counter-movement to Christianity, which will gain enormous power as mankind passes into the final phase of its earthly existence. What confronts us here is not simply an "axis OX" set at right angles to "the Christian axis OY," but a hostile Force of supra-human proportions and a final battle to the

---

41. Luke 16:13.
42. Matthew 10:22.

death. It appears moreover that this counter-movement will be realized in the form of a collective human organism replete with its own Antichristic "center of attraction": it will in all respects be a caricature, a kind of satanic imitation or inverted image of the true Mystical Body of Christ. And this super-organism will grow and wax great by deceiving vast multitudes with its clever lies and marvelous feats. And in the end it will hoist religious colors and proclaim itself Divine.

It is hardly surprising that Teilhard has little to say on that subject and seems to avoid it like the plague.

Dietrich von Hildebrand relates how in the course of a conversation with Teilhard de Chardin he happened to refer to St. Augustine. "Don't mention that unfortunate man" Teilhard exclaimed instantly; "he spoiled everything by introducing the supernatural."[43] To which one might add that long before the Bishop of Hippo, Christ had obviously committed the same offense when He declared: "My kingdom is not of this world." It is not often, however, that Teilhard exhibits his "crass naturalism," as von Hildebrand calls it, quite so openly.

What Teilhard is telling us, basically, is that all forms of "progress"—and most especially our scientific and technological advances—are actually taking place under the aegis and inspiration of Christ, unacknowledged though this fact may be, and that all these scientific and technological developments are contributing to the formation of the Mystical Body, which is in fact coextensive with the emergent social organism.

Now this tenet constitutes one of Teilhard's crudest mistakes. Ostensibly he bases his conclusions upon the presumed universality of Christ: if Christ is universal—so the implicit argument goes—then His Kingdom can have no bounds. But what Teilhard fails to grasp is that whereas Christianity maintains that the whole of humanity has indeed been redeemed by the Risen Saviour, the fact

---

43. *Trojan Horse in the City of God* (Chicago: Franciscan Herald, 1967), p. 227.

remains that everyone is free to participate in that Redemption or to exclude himself therefrom. In a word, we are saved potentially—"for the asking," if you will—but we are not yet actually saved: there is a crucial choice to be made, to say it again. We begin to participate in the supernatural life—we "receive the Holy Ghost"—as soon as we have done our part, but not a moment before. What is needed is a conversion, a catharsis, and an initiation—and we may leave it to the theologians to spell out more precisely what this entails. There is in any case a decisive step that must be taken, consciously and with deliberation, before one may become a *bona fide* member of the Mystical Body, also known as the Church.

The universality of Christ, therefore, and of the Redemption which He has won for mankind upon the Cross, must not be interpreted to mean that the Mystical Body is coextensive with humanity at large, or that it will be such in the future. Christ Himself has made this perfectly clear: for instance, when He speaks of the good grain and the tares growing together in the same field.

So too the supernatural life begins here and now for those who are prepared to receive the gift: we need not wait for death and Judgment to occur. Through baptism and all that by right pertains to this sacrament we are born into a new life: we "put on Christ" as St. Paul declares. And that new life cannot be measured in terms of outer forms, which is to say that a Christian is able to live his Christianity not only in prayer and contemplation, but in virtually every worthy activity. St. Paul himself continued to make tents, and no doubt transformed that métier into a profoundly Christian occupation. And this is just the point: an action which ordinarily is profane can be Christianized, so to speak, and thus taken up into the supernatural life, which is the life in Christ. It is then profane no more, nor is it still a merely natural act. An infusion of grace has occurred, a descent of the supernatural—a transfiguration, one could almost say. But this does not mean that the line of demarcation between the Kingdom of Christ and the material world has been blurred or obliterated: it does not mean that near its outer fringes, so to speak, the former ceases to be strictly supernatural. To entertain such a view is to misunderstand completely the meaning of the Christian life.

In Teilhard's theory salvation is perceived as the ultimate product of socialization combined with a fantastic technology. But this is a colossal mistake: mankind is to be transformed into a super-humanity, not by some linkage with itself through technical means, but by a union with God through *spiritual* means. *Deo volente*, we shall all be one in the Spirit of God. But that Spirit is not something that has evolved, or that shall evolve, as Teilhard maintains. The problem is not to *create* that Spirit—who in fact is the uncreated Creator of us all—but simply to receive the divine Gift, to become receptive. And this is a matter, not actually of the brain, but of the heart: it is "the pure in heart" that shall see God: "Verily, I say onto you, Whosoever shall not receive the kingdom of God as a little child, he shall not enter therein."[44]

Teilhard is hopelessly mistaken when he thinks that the authentic spiritual life will be achieved through some kind of super-cephalization: nothing could be further from the truth. The brain, after all, is only an instrument; it is a computer, if you will; and most assuredly, no one has ever become spiritual through some stupendous calculation. There is something else in us—call it soul, intellect or spirit—and this is what needs to be purified and awakened. If the sanctified man can "see God," it is certainly not by way of cerebral activity! What ultimately counts is spiritual sight, a knowing that is no longer conceptual, no longer indirect: "And this is life eternal, that they might know thee the only true God, and Jesus Christ, whom thou hast sent."[45]

But needless to reiterate, the brain has nothing whatsoever to do with this consummate knowing. It has to do, rather, with the outer life: our life in *this* world. It is an instrument, one can say, adapted to this particular sphere of existence, which is needed precisely so long as we are *not* in the Spirit.

Now, one might perhaps object to the aforesaid on the grounds that Christianity speaks, after all, of a Resurrection. But does this dogma imply that in the life to come we shall possess a brain made up of neurons? Certainly not! What it does imply is that this earthly brain made up of neurons manifests a spiritual prototype—an Idea,

---

44. Mark 10:15.
45. John 17:3.

if you will, in the Mind of God—and can consequently be realized or exemplified in the spiritual world. The body can be spiritualized, can be transfigured, precisely because it is *not* just an aggregate of particles. It enshrines something that survives the dissolution of its particles: that is what the Christian dogma of the Resurrection entails. And this means also that, even here and now, the particulate brain is neither the knower nor the knowing but only an instrument.[46]

Getting back to the Mystical Body of Christ: Scripture teaches that this Body came into being through a supernatural influx of the Holy Ghost: "I came to cast fire upon the earth...."[47] It came into existence, therefore, "from above" and not "from below" as Teilhard's theory implies. And what is more, it came to be *suddenly*, in an instant, as befits the action of the Holy Ghost: "And suddenly there came a sound from heaven as of a rushing mighty wind, and it filled all the house where they were sitting."[48] Nothing could in fact be further from the Christian truth than the notion that the Mystical Body is the product of an evolution extending over millions of years.

Now the Body that came to birth at Pentecost is indeed a supernatural or spiritual organism. It belongs to a different plane, and constitutes in fact a new creation, a new world. Strictly speaking, the Mystical Body is none other than "the World to Come": it is the New Jerusalem, the Kingdom of Heaven. And yet it has entered into this world, has entered into history. It is not a thing of the future any more, as it was in the days of the Old Testament, but a present reality. That is the great mystery: "The time is fulfilled and the Kingdom of God is at hand."[49] Whether we realize it or not, the New Jerusalem is already standing before us; it has penetrated into our world and we are truly living in "the latter days." The old order, the purely "natural" world, is still there, to be sure, but it is standing on its last legs: its days are numbered. For even now the Mystical

---

46. On this issue I refer to my chapter on "Neurons and Mind" in *Science and Myth* (Tacoma, WA: Sophia Perennis/Angelico Press, 2012).
47. Luke 12:49.
48. Acts 2:2.
49. Mark 1:15.

Body—the new creation which is destined to supplant and supersede the old—is rising up in its midst.

This is the veritable "super-organism" of which Christianity speaks; and let us be careful to add that it is indeed supernatural. Though *in* the world, it is yet separated therefrom: like soul and body, the two do not mix. For all its proximity, the Mystical Body is by no means continuous with the natural order, but is situated, as one might say, on a higher ontological plane It is therefore invisible to our natural organs of perception, and to the profane intelligence which is geared to the realities of the sense-perceived world: "Flesh and blood has not revealed it unto you...."[50]

The Mystical Body is not produced by human means as Teilhard would have us believe, but is rather God's free gift to mankind. This "super-organism" is moreover an accomplished fact: it stands before us here and now. We are not obliged to set about with Promethean prowess to construct the New Jerusalem; the one thing needful, rather, is to make ourselves receptive, to become humble and docile before God. All Christ asks of those who would enter the Kingdom of Heaven is that they follow Him who is "the Way, the Truth, and the Life."[51]

Such is the calling; and let us understand that it is addressed, not to society at large, but to men and women of every description. It is Peter and Paul—and not some human conglomerate—that is to be incorporated into the Mystical Body. And this happens, when it does, not after so many centuries, but in the twinkling of an eye—because the miracle is wrought, not by human endeavor, but through the operation of the Holy Ghost.

There are, of course, degrees of adhesion to the Mystical Body of Christ. Yet it remains true that a certain knowledge imparted by the Holy Spirit is indeed "the hallmark and seal of the believer" as Archimandrite Vasileios has said. Every true Christian has at times

---

50. Matthew 16:17
51. John 14:6.

experienced—however faintly—the "flavor" of the life in Christ, and the "taste" of holiness. Whether he realizes it or not, he exists already on two planes and knows two lives; but the worldly-minded experience only one. When we grow into manhood we do not cease to understand the pastimes of children; we only lose our relish for these erstwhile pursuits: we have no more desire to engage in these once-fascinating games because we have found something better. There is an inherent asymmetry between the two states which betokens a change of level, a change of plane.

The Christian can very well live in a technological society, and can even participate in its projects. Yet inwardly he remains aloof, for his heart is set on a very different goal. And when it comes to the satisfaction of his deepest aspirations he has no need for secular institutions, be they research institutes, industries or what have you: even our fantastic means of communication cut little ice when it comes to that. To put it very simply: the authentic Christian experiences a daily inclination to withdraw into the solitude of his heart, that inner "closet" Christ bids us enter, to commune with God and drink to his heart's content of the spiritual nectar. And having done so—having become filled and fortified with the gifts of the Spirit—he descends, as it were, into the humdrum world to share of his bounty and serve his neighbor who is in need. Such is the fully Christian life which we should all strive to attain. It is a life lived on two planes; or better said, it is a heavenly life lived here and now on this earth.

One finds that it is in truth the men of this world, inclusive of those who are supposedly engaged in the laborious task of constructing the Teilhardian "super-organism," who are childish and misguided. They seek Heaven in the building of some gigantic Babylon, not knowing that Heaven is already near at hand, that in fact it lies within easy reach. What a pity! What a tragic farce.

Meanwhile the Mystical Body is daily extending its domain. It is forming in the quiet of deserts and mountain caves, and amidst the tumult of great cities. It sprouts and bursts into bloom wherever a human heart loves Christ. We have argued earlier that on the whole our modern preoccupation with science and technology is not congenial to spiritual growth, and that may be true. Yet the fact remains that authentic spiritual life is possible everywhere; the Spirit, like the wind, "bloweth where it listeth"—even in factories. All that is needed is a human heart that has not yet been killed, and some living contact with the Mystical Body of Christ which is the Church. These are the two requisites; and apart from this nothing is needed—nothing at all.

10

THE NEW RELIGION

FOLLOWING THE PUBLICATION of Darwin's magnum opus it was generally assumed in evolutionist circles that religion *per se* had become an outmoded superstition, and the expectation was rife that in time these "primitive vestiges" would give way before the advancing front of scientific enlightenment. There were still believers, to be sure, and there was yet the Old Church rising up like a medieval fortress within the modem world. But the stronghold was obviously under siege, and by human reckoning it seemed that its days were numbered.

This is where Teilhard de Chardin enters upon the scene. He proceeds forthwith to develop a position of his own, beginning with the simple recognition that science and neo-humanism in its various manifestations constitute in essence a single integral movement of worldwide scope. It appears to him, moreover, that there is something distinctly religious about this contemporary movement: "A religion of the earth is being mobilized against the religion of heaven"[1] he declares. And this supposition leads to a third step: having diagnosed that a single neo-humanist "religion" of global proportions is being marshaled against "the religion of heaven"—by which Teilhard obviously understands Christianity—he goes on to conclude that this new "religion of the earth" can be nothing more nor less than the manifestation of the evolutive thrust on the collective human plane. And from this presumed discovery he naturally concludes—apparently with all his heart and soul—that the movement in question constitutes in fact the one and only true religion, a

---

1. SC, p. 120.

conclusion which leads finally to the last turn in this Teilhardian dialectic. It is actually the only option left to him as a member of the Catholic Church: he declares apodictically that Christianity and neo-humanism are pointing in the same direction and must henceforth join forces; in a word, there is to be a New Religion.

It appeared that Teilhard had deftly turned the tables on the evolutionists of the old school: religion—and indeed Christianity no less—so far from having become a thing of the past, is to be *the thing* of the future. No wonder loud cheers and bursts of applause could be heard rising up from behind the ramparts of the Old Church! Nor is it too surprising that ere long some gates of the Fortress came to be opened from within.

But the question remains of course whether the proposed merger between Christianity and the neo-humanist movement is legitimate. It may be comforting to think that the Church is no longer under siege, and that her erstwhile opponents are now ardent champions of her cause. And for many who had been teetering in their faith, the notion that science itself guarantees the essentials of Christian dogma must have come as a blessed relief. But needles to say, exhilarating as all this may be, such euphoria proves nothing. Profound theological questions stand at issue, and Huston Smith may yet be right when he observes that "only an exhausted theology, one about to sink into the sands of science like a spent wave, could fail to sense the enormous tension between its claims and those of a scientific world view."[2]

One needs to ask oneself whether the Teilhardian scientization of Christian belief may not in fact be destroying the very thing which it is supposed to save and perfect. Now it appears that Teilhard himself experienced at least some momentary qualms on that score. In one of his earlier letters, for example, he admits that "sometimes I am a bit frightened to think of the transposition to which I have to subject the vulgar[3] notions of creation, inspiration, miracle, original sin, resurrection, and so forth, in order to be able to accept

---

2. *Beyond the Post-Modern Mind* (NY: Crossroads, 1982), p. 108.
3. Dietrich von Hildebrand may be right when he observes that this use of the term "vulgar," though perhaps not meant in a pejorative sense, is indicative of a

them."⁴ Yet it seems that before too long Teilhard was able to cast off these trepidations and become reconciled to the idea that he was actually founding a new religion. He makes it clear to a chosen few that such is indeed his intention; in a letter to Leontine Zanta, for instance, he writes:

> As you already know, what dominates my interest and my preoccupations is the effort to establish in myself and to spread around a new religion (you may call it a better Christianity) in which the personal God ceases to be the great neolithic proprietor of former times, in order to become the soul of the world; our religious and cultural stage calls for this.⁵

It thus turns out in the end that Teilhard is not after all the champion of a besieged Christianity, but rather the founder of a new religion supposedly destined to supplant the old. Despite his contention that "you may call it a better Christianity," it happens that the new cult is not the Christianity of bygone days. In fact, it is so radically different that Teilhard refers to it as "a hitherto unknown form of religion—one that no one could as yet have imagined or described, for lack of a universe large enough or organic enough to contain it."⁶ Not only, then, was it nonexistent in ancient times, but it would not have been possible even to conceive of that new religion in a pre-scientific and pre-Darwinist age. And as if this were not enough, Teilhard adds by way of further clarification that the new religion "is burgeoning in the heart of modern man, from a seed sown by the idea of evolution."⁷ But if such be the case, not only then was the new religion unimaginable in olden times, but it did not even originate from the ancient stock but sprang instead from "a seed sown by the idea of evolution." In plain terms, the new religion derives, not from

---

"Gnostic" outlook. See *Trojan Horse in the City of God* (Chicago: Franciscan Herald Press, 1964), p. 239. See also my article "Gnosticism Today," *Homiletic and Pastoral Review*, March, 1988.

4. Letter dated December 17, 1922, quoted by Philippe de la Trinité in *Rome et Teilhard de Chardin* (Paris: Fayaard, 1964), p. 47.

5. *Lettres à Leontine Zanta* (Paris: Desclée de Brouwer, 1965), p. 127.

6. AE, p. 383.

7. Ibid.

Yahweh or Christ, but yes, from the author of *The Origin of Species* himself! One cannot but wonder whether Teilhard was playing square with us when he declared that "this is still, of course, Christianity."[8]

In retrospect one sees that those shouts of jubilation from behind the ramparts were premature, and some of those, perhaps, who unbolted gates are having second thoughts.

It would however be a grave mistake to write the impact of Teilhardism off as a passing fad or a momentary madness. So far as the Church in most parts of Europe and America is concerned, the influence of the French Jesuit has undoubtedly been deep, pervasive, and lasting. It is of course difficult to distinguish the impact of Teilhardian beliefs from that of the Council. The decade of the sixties constitutes one of those moments in history when change of one kind or another was in the air, and in its outer manifestations the Church has never since been the same. And in certain respects and to some degree it has in fact turned Teilhardian. One can see this, for instance, in such phenomena as the radical involvement on the part of the bishops in political and economic issues (almost invariably left of center), the waning of faith in the supernatural, and the ongoing deconversion of clergy and laity alike from all "staticist" beliefs. The trend is unmistakable: Christianity as personified, in the first place, by major contingents of the Roman Catholic hierarchy, as well as by various Protestant and inter-denominational institutions such as the World Council of Churches, has begun to turn in the direction mapped out by Teilhard de Chardin. Whether or not these segments of official Christendom have by now "embraced in love" what Teilhard terms "that tremendous movement of the world which bears us along,"[9] it is clear in any case that a concerted effort to merge the two "religions" is well under way.

This state of affairs has of course been often enough observed,

8. HM, p. 96.
9. PM, p. 298.

and not a few historians of religion have recognized the Teilhardian connection. As Albert Drexel, a Catholic ecclesiastic, explains with admirable precision:

> The modernism or neo-modernism within Christianity, and especially within the Roman Catholic Church after the Second Vatican Council, is above all characterized by a turning away from the supernatural and an exclusive predilection for this world, the Aggiornamento of Pope John XXIII interpreted one-sidedly and hence misapplied. Teilhard's ideology was a definitive precondition for this. Inasmuch as he turned his back to the past, fused God and the supernatural with the process of a universal evolutionism, and proclaimed religion to be an active participation in a progressive development ending in Point Omega, the basis was given for a humanist cult of the secular ("ein humanistischer Diesseitskult").[10]

Given that such is the trend, and having identified its "definitive precondition," it remains to ask what has been its effect upon the religious life of the faithful. And surprisingly, the answer to this delicate question was given in 1984 by the top theologian at the Vatican: the former Cardinal Ratzinger, namely, who later became Pope Benedict XVI. "The results were totally opposite to the hopes of all" admits the Cardinal in an interview.[11] "We had hoped in renewed Catholic unity, and the results have been a pattern of autocriticism leading to self-destruction. We had hoped for a new enthusiasm, but the results have been discouragement and boredom." He concludes that the post-Conciliar period has been "decidedly negative for the Catholic Church." In place of the "renewal" we have heard so much about, the Cardinal now speaks of a pressing need to "restore the Church." It is clear that Cardinal Ratzinger does not believe in a marriage of Teilhard's "two religions." In fact he declares his stand on this crucial issue with the utmost precision and emphasis: "He is totally ignorant of the nature of the Church and of the nature of the

---

10. *Ein Neuer Prophet?* (Stein am Rhein: Christiana Verlag, 1971), p. 127.
11. The interview has been published in full as *The Ratzinger Report* (San Francisco: Ignatius Press, 1985).

world," says Ratzinger, "who believes that these two can meet without conflict or that they may be somehow mixed." And he urges bishops and ecclesiastic leaders to change their course and oppose "the many worldly cultural tendencies adopted by post-Conciliar euphoria."

Whether or not Cardinal Ratzinger's voice—which has now become the voice of Pope Benedict—will prevail is of course an open question. Yet it is in any case significant that twenty years after Vatican II such an assessment should have issued from the highest levels of the Vatican itself.

Getting back to Teilhard's self-assumed mission, it is to be noted that the idea of a new religion was never prominently displayed in his voluminous writings. Certainly he did not hide the fact that major changes had been proposed; yet he generally conveyed the impression that his sole objective was to inaugurate a more highly evolved Christianity. In an evolutive universe, so the argument goes, how can the Church stand still? At one point he goes so far as to characterize his mission as "the laying of new foundations to which the old Church is gradually being moved,"[12] which however leaves open the crucial question whether the newly-founded structure will still be the Church.

To begin with, it is a strange notion that the Church *can* be "moved to a new foundation," considering that it had always been understood that these foundations have been established, once and for all, by God Himself: "Thus saith the Lord God, Behold, I lay in Zion for a foundation a stone, a tried stone, a precious corner stone, a sure foundation."[13] So too St. Paul—whom Teilhard delights to quote in other contexts—declares explicitly: "For other foundation can no man lay than is laid, which is Jesus Christ."[14] Nor must it be supposed that this everlasting Foundation is simply Jesus Christ

---

12. FM, p. 23.
13. Isaias 28:16.
14. 1 Corinthians 3:11.

conceived *in abstracto*, so to speak, as if prophetic and apostolic tradition derived from Him had nothing to do with that stipulated Foundation. St. Paul himself makes this quite clear when he speaks of the living Church as "built upon the foundations of the apostles and prophets, Jesus Christ himself being the chief corner stone."[15]

So long, therefore, as we take seriously the word of Scripture, the question of moving the Church to new foundations has been set at rest once and for all. But what if we don't? As we have had ample occasion to note, Teilhard is always willing to take Biblical teachings *cum grano salis* and jettison those parts which do not harmonize with his preconceived notions. Now that in itself, to be sure, makes him a heretic according to the classical definition of the term. But perhaps that concept, too, needs nowadays to be revised or discarded. Perhaps, in light of Evolution, we need now to abandon the old "staticist" views and adapt our religious outlook to the newly-revealed facts, beginning with the so-called "discovery of Time and Space." And so we come back once more to the notion of moving the old Church to new foundations.

From a humanist point of view, the Church, like any other institution, can of course be revamped at will whenever it may seem expedient to do so; and as to the question whether "this is still, of course, Christianity," it could be argued that we are free, after all, to call the revised cult or religion whatever we please. Words are cheap in this nominalistic age.

Now the first thing to be noted in regard to such arguments is that Teilhard's position can indeed be attacked and refuted, but *not* on the basis of theological premises or metaphysical principles, the validity of which he radically denies. To be cogent and effective, the refutation of Teilhardism must first be carried out on Teilhard's own presumed turf, that is to say, on *scientific* ground. We must remember that time and again Teilhard has made it a point to disavow both theology and metaphysics—which permits him, in effect, to offend against both disciplines with impunity, provided only that he can convey the impression of speaking as a scientist. He argues that the Church must be "moved" because its erstwhile teachings do

---

15. Ephesians 2:20.

not harmonize with the newly-discovered truths of science. But it happens that this claim is *demonstrably* false, as we have in fact had ample opportunity to see. From a strictly scientific and indeed profane point of view it can be said with certainty that *science as such neither demands nor can authorize the revisions for which Teilhard pleads in its name.*

But there is something else that needs also to be noted: when it comes to the Church and its doctrines, the profane point of view does not reach very far. The real Christianity is not to be known from the outside: "flesh and blood" do not suffice for the discovery of that truth. Its doctrines cannot actually be examined and weighed by unbelievers; what Philip said to Nathaniel applies to us all: "Come and see!"[16] We must bestir ourselves and step away from the "fig tree" of this world, and like Nathaniel, we need to be "without guile": then alone can we know what Christianity is about. And then shall we realize, strange as it may sound, that the Church lives not so much in time as in eternity: for it lives in truth by the Holy Spirit. In a real sense, therefore, the Church does not change at all. It may of course change in some of its outer manifestations, but never in its essential truth and innermost life; as Georges Florovsky has well said: "In the life and existence of the Church time is mysteriously overcome and mastered; time, so to speak, *stands still.*"[17] Our task, then, as Christians, is not to recreate the Church by tampering with its foundations, but to drink ever more deeply from its sacred vessels and savor the timeless draught. And when we shall have received the Holy Spirit as we should, we will no longer perceive the Church in temporal or "evolutive" terms, but indeed *sub specie aeternitatis.*

Teilhard does not always speak of shifting the old Church to new foundations; more often than not he speaks as one who would bring the Church back to its essential truth. For example, in an

---

16. John 1:46.
17. *Collected Works* (Bellmont, MA: Nordland, 1972), vol. 1, p. 45.

essay entitled "Introduction to Christianity" he takes it upon himself to exhibit the essentials of Christian dogma in the form of three successive articles of faith. They are as follows:

1. Faith in the (personalizing) personality of God, the focus of the world.

2. Faith in the divinity of the historic Christ (not only prophet and perfect man, but also object of love and worship).

3. Faith in the reality of the Church *phylum,* in which and around which Christ continues to develop, in the world his total personality.[18]

"Apart from these three fundamental articles," Teilhard informs us, "everything else in Christian teaching is basically no more than subsidiary development or explanation (historical, theological, ritual)."

Now it is true, certainly, that each of the three articles (despite linguistic quirks) admits of an orthodox interpretation. But what about Teilhard's contention to the effect that this meager catechism comprehends all that is essential in Christian doctrine? Gone, first of all, is "God, the Father Almighty, creator of heaven and earth"; gone, too, is the Holy Spirit; and in the realm theologians are wont to designate "*ad extra,*" gone is the Resurrection of Christ, the communion of saints, the forgiveness of sins, and numerous other things besides. Are all these omitted articles of belief then inessential to the Christian faith? Is the Resurrection of Christ, in particular, to be dismissed as a "subsidiary development or explanation"? Just a moment ago Teilhard had quoted the verse from First Corinthians (one of his favorites) which reads: "And when all things shall be subdued unto him, then shall the Son also himself be subject onto him that put all things under him, that God may be all in all"; but he does not seem to stand with the Apostle when the latter declares in the same chapter: "If Christ be not raised, your faith is vain."[19]

It is unrealistic, furthermore, to suppose that a list of statements

---

18. CE, p. 152. My italics.
19. 1 Corinthians 15:28 and 15:17, respectively.

(of whatever length) could enshrine the essentials of Christian doctrine, nor is this actually what a catechism is meant to do: for it is only within the living tradition of the Church that such "dogmatic definitions" can be rightly understood. Those, therefore, who have cut themselves off from that tradition, or who have never participated therein, are not apt to be enlightened by such formal means. And whereas even the most orthodox formulations of Christian dogma can obviously be misinterpreted, when it comes to the Teilhardian "articles" a heterodox interpretation is virtually assured by the fact that Teilhard has deftly installed a few "signposts" of his own. Taken within the context of the Teilhardian doctrine, as they are meant to be, these artful statements amount to no more than a reaffirmation of Teilhardian tenets. Thus "correctly" interpreted, the first article, for instance, reduces evidently to the affirmation of Point Omega: no more and no less.

But what about the second article ("Faith in the divinity of the historic Christ"): what could be more orthodox than that? Yes, but what exactly does Teilhard have in mind when he speaks of "divinity"? Faith in "the divinity of the historic Christ": what can this possibly mean under the auspices of Teilhard's theory? The question is not entirely simple, and that is perhaps the reason why Teilhard comes to our aid. He has provided us with a hint placed in parentheses: "not only prophet and perfect man, but also object of love and worship." This is still a bit vague, of course; and yet the suggestion is unmistakable: the divinity of Christ is actualized through the love and worship accorded him by mankind. He is divine, in other words, insofar as he is actively fulfilling his role as "the focus of the world." We need not be surprised. Has not Teilhard told us time and again that henceforth mankind can believe in no other God save Point Omega?

Regarding the third article of the Teilhardian catechism, suffice it to observe that in speaking of "the Church *phylum*" (a funny phrase), Teilhard is dropping a broad hint to the effect that the Church, like everything else, has evolved, and constitutes at bottom a biological phenomenon. It is only that our Christian forebears were too primitive to recognize the fact; they still believed in the supernatural, and in an "extrinsicist" God "whom no one in these days any longer wants."[20] It is one of Teilhard's favorite themes, to which he returns time and again: "For hundreds of centuries (up to yesterday, one might say)," he tells us, "men have lived as children, without understanding the mystery of their birth or the secret of the obscure urges which sometimes reach them in great waves from the deep places of the world."[21]

It could hardly have been put more plainly: "up to yesterday"—until Darwin came along, that is—men were unable to comprehend the source and true meaning of their deepest aspirations. The sages and prophets of old were mistaken when they interpreted the mystic urge within their soul as the beckoning of God, the response of the human heart to His call. They did not realize that such aspirations are in reality biological, that they spring "from the deep places of the world."

But let us get back to Teilhard's abbreviated catechism. Having singled out his "three fundamental articles" (replete with evolutionist guideposts) and turned his back on Christian tradition, which by his own reckoning is to be ranked among "the whims and childishness of the earth,"[22] Teilhard is free at last to unveil the new theology; and it is hardly surprising that at his magic touch everything acquires a new face. Let us consider a few examples.

Speaking of "grace" Teilhard has this to say: "From the Christian, Catholic and realist point of view, grace represents a physical supercreation. It raises us a further rung on the ladder of cosmic evolution. In other words, the stuff of which grace is made is strictly biological."[23] Let us try to enter into this remarkable train of thought. The first statement is more or less orthodox, except for the adjective "physical," whose meaning in this context is not initially clear. Presumably that adjective is there to guide us into the universe of discourse in which the second statement is to be understood: we are now in the domain of evolutionist thought, in which the "supercreation" of Christianity becomes simply "a further rung on the lad-

---

20. FM, p. 279.
21. HE, p. 32.
22. Ibid.
23. CE, pp. 152–153.

der of cosmic evolution." And Teilhard loses no time to draw the desired conclusion: "In other words" he goes on to say, "the stuff of which grace is made is strictly biological." With seeming logic Teilhard has taken us from an isolated proposition, which might sound Christian enough, to a statement concerning the nature of grace which affirms very much the opposite of what Christians had always believed: for it belongs to the very essence of grace to be, not a natural attainment, but a supernatural gift. What Teilhard has actually done under the pretext of interpreting the term is to deny that such a thing as grace exists. His "grace" is actually a "non-grace," if one may put it thus.

Or consider what befalls the conception of infallibility: "In reality," Teilhard explains, "to say that the Church is infallible is simply to say that, in virtue of being a living organism, the Christian group contains in itself, and to an eminent degree, a certain sense of direction and certain potentialities: ill-defined though these are, they, enable it to grope its way, constantly probing in this direction or that, to maturity or self-fulfillment."[24] Now admittedly it is not easy to formulate the notion of ecclesiastic infallibility in precise terms, and there is doubtless room for differences of interpretation. But one need hardly be a theologian or a canonist to realize that the stipulated infallibility derives from the fact that the Church is not just a "living organism," but indeed a *theandric* organism, which entails that this infallibility derives ultimately from the mystery of the Hypostatic Union. Or to put it another way: the Church is infallible in a certain sense because, in a distinctly supernatural way, it stands under the inspiration and guidance of the Holy Spirit. But of course all this transcends categorically the confines of Teilhard's *Weltanschauung*. And so, once again, Teilhard's pretended definition turns out to be in effect a denial that the thing in question exists; for if indeed the Church is infallible "in virtue of being a living organism," then we arrive at the strange conclusion that this infallibility is shared by amebae and buffaloes! And since we can presumably all agree that amebae and buffaloes are not in fact infallible, it follows that neither is the Church.

24. Ibid.

Getting back to infallibility in the authentically Christian sense, let us also note that this conception does *not* apply to the sphere of action: it has nothing whatsoever to do with "groping one's way towards maturity and self-fulfillment," but applies to the doctrinal sphere. By misrepresenting the idea of infallibility as a pragmatic wisdom of some kind, Teilhard has at the same time denied by implication the universal validity of dogmatic formulations: so long as the notion refers to a self-corrective groping, something to be found in the behavior of animals or even of servo-mechanisms, it obviously cannot apply to such things as the Christological affirmations of the Councils.

It will hardly be necessary to follow Teilhard further as he proceeds to explain, in succession, the Trinity, the divinity of Christ, Revelation, Miracles, Original Sin and Redemption, Hell, and finally the Eucharist. It is everywhere the same story: all that exists in the end is the evolutive process: "That, when all is said and done, is the first, the last, and the only thing in which I believe."[25]

It is literally true that Teilhard has deified Evolution. And from the start the concept of evolution had been—not just a scientific hypothesis, nor even a scientifically established fact—but an idea charged with a kind of religious significance. Teilhard de Chardin was presumably the first person in history to be totally possessed by the concept, the first to be fully intoxicated with the new wine. It seems that Darwin himself was still to some extent rooted in the past, not yet completely "liberated." Not until Teilhard appeared upon the stage, at any rate, did Evolution find its full-blown prophet. It was he that brought out—with a fury one can say—the religious pretensions which presumably had been latent in the movement from the start. At Teilhard's hands the Darwinist theory became transformed into a full-fledged religion: was actually turned into a cult.

This explains "that tipsy, euphoric prose poetry," and those

25. Ibid., p. 98.

"alarming apocalyptic seizures"[26]: such things are nothing unusual in a man fired with a sense of religious mission. Teilhard is convinced he sees, not just for himself, but for all mankind—nay, for the cosmos at large! At his best he does not write: he cries out in a loud voice. But unlike the prophets of old, his is not the voice of Tradition: on the contrary, it is manifestly the voice of anti-Tradition: "A new victorious passion is beginning (we seriously believe) to take shape, which will sweep away or transform what have so far been the whims and childishness of the earth"[27] he cries.

What makes this all the more ominous is that the new cult sports Christian colors. One cannot say "masquerades in Christian garb," because it is all too obvious that the new religion is diametrically opposed to the old on just about every count. What has happened, as we have just seen, to the ideas of grace and infallibility is by no means atypical: it turns out to be the rule. And what else could one expect once it has been admitted that "we no longer want a religion of regulation: but we dream of *a religion of conquest!*"[28] Even the *Pater Noster* has become turned around: henceforth it is no longer "Thy will" but *our* will that is to be done. The fact is that Teilhard stands on the side of that "religion of the earth" which according to his own account "is being mobilized against the religion of heaven." Yet even so he persists in the blatant claim that "this is still, of course, Christianity." And with telling effect: for as the exiled Jesuit had shrewdly foreseen, the newly-hatched anti-creed has come to be accepted by millions as the true Christianity. In the eyes of the true disciple it is every whit the ultra-Christianity Teilhard declared it to be.

It is true that Teilhard preaches Jesus Christ. But on what basis can an evolutionist do so? Multitudes are impressed and fascinated when a man of science proclaims Christ to be the focus of the world.

---

26. We are again quoting Peter Medawar.
27. HE, p. 32.
28. SC, p. 103.

They do not know, first of all, that the so-called "Omega Point of science" is a fake. So too the question remains how someone who has rejected tradition—who thinks that until yesterday "men lived as children"—could know that Jesus of Nazareth became Point Omega. We know Jesus almost entirely through Christian tradition; and if we reject that tradition and its claims, what remains could hardly suffice to buttress such a gigantic assertion. How, then, can Teilhard speak of Jesus Christ as the universal center of attraction? As von Hildebrand has very aptly pointed out: "An unprejudiced mind cannot but ask: Why should this 'cosmic force' be called Christ?"[29] Why indeed? Is it because Jesus Christ enjoys a certain "popularity" in our world? But then, so does Buddha, and so does Mohammed; and so, too, do many secular figures upon the world stage.

We need not belabor the point: from a scientific perspective there simply *is no* connection between Jesus of Nazareth and Point Omega. It is not enough to say that there is no evidence for such a claim: for even if a Point Omega did exist, the Christological connection would still be, not only unverified, but scientifically inconceivable. The fact is that *only* by way of the Christian tradition can we begin to know Christ, and only on that basis could He be recognized as the true Center of the world. But once that tradition has been undercut and denied, what then? If it be true that our Christian forebears were infantile and their religious beliefs mere "whims and childishness of the earth," on what basis can Teilhard preach Christ? Clearly, *there is no such basis*: this too is a fake.

Despite his clerical garb, Teilhard's attitude towards traditional Christianity is distinctly negative and critical. It is not clear whether there is anything at all in the old religion of which he approves. At times he does of course extol certain Christian conceptions, for instance the idea of personalization. Yet on closer examination one invariably finds that the praises he sings actually apply, not to the old, but to the new "Christianity."

---

29. Op. cit., p. 238.

Teilhard never misses an opportunity to criticize and discredit whatever is not to his liking or stands in his way. Nor does he hesitate to cast aspersions even upon the words of Christ Himself. Even the Beatitudes, for example, do not escape the ire of his reformist zeal: "There can be no place for the poor in spirit," and no place either for "the sad of heart"[30] he exclaims.

His central complaint, to be sure, is that traditional Christianity is not "scientific," by which he evidently means, first and foremost, that it does not harmonize with the evolutionist outlook It is therefore "staticist" and must be revised.

A second objection, moreover, closely related to the first, is that the old beliefs are not compatible with our so-called "discovery of Time and Space." Now on that score Teilhard may have a point. When it comes to the physical dimensions of the universe—and more generally, to the material sphere in the specifically modern sense of that term—it can hardly be denied that our forebears may indeed have been somewhat naïve. But Teilhard forgets that numerical or physical magnitude—the immensity of Avogadro's number or of the Hubble constant—is not everything. There is also a spiritual immensity or "*qualitative* plenitude" if you will; and that is after all the true immensity: it is the plenitude of being and of life itself, which is infinitely greater than the enormity of its outer shells.[31]

If the Weltanschauung of Christianity seems narrow to Teilhard de Chardin it is doubtless due largely to the fact that he has never approached the subject without preconceived notions of a scientistic kind and the hubris of someone who imagines he has outstripped all who came before.

It is quite apparent that Teilhard does not resonate to the Biblical world-view. Not only does he find it "staticist" and narrow, but he

---

30. FM, p. 75.
31. I have argued repeatedly that the so-called physical universe, rigorously conceived, needs to be distinguished from the corporeal world which we behold with our senses, and that even physics on its most fundamental level (i.e., quantum theory) cannot be interpreted ontologically without recognizing that distinction. See *The Quantum Enigma* (Tacoma, WA: Sophia Perennis/Angelico Press, 2008). As regards the wider implications of this recognition I refer to *Science and Myth*, op. cit.

thinks it is based upon "the Alexandrine" cosmology (whatever that might be). What Teilhard fails to grasp, however, is that the Biblical *Weltanschauung* is actually based, not upon any theory or abstract model of the universe, but quite simply upon sense perception. The Bible speaks of the world as it is revealed to us—not by way of Geiger counters or telescopes—but through our God-given organs of perception. Christianity has never claimed, moreover, that this perceived world is absolutely real: "Now we see as through a glass, darkly."[32] But nonetheless "the invisible things of Him from the creation of the world are clearly seen, being understood by the things that are made, even His eternal power and Godhead" as Saint Paul likewise declares.[33] The heavens above and this solid earth beneath our feet—everything without exception is endowed with boundless significance: to the wise all things in creation point in fact to God. For those who *see* the world is a holy icon; and this fact both explains and justifies the Biblical world-view.

Actually the shoe is on the other foot: it is Teilhard who is duped. It is he that has mistaken a mere abstraction, a mere "model," for the reality itself, and has thereby succumbed to what Whitehead terms "the fallacy of misplaced concreteness." Let those who disavow philosophy stay clear of the despised discipline!

Another frequent Teilhardian complaint pertains to the so-called "juridical" notions of traditional Christianity. Now it may well be true that there has been in later times a tendency in certain quarters to overemphasize this aspect of Christian doctrine. But Teilhard should have been cautioned by the fact that juridical ideas do obviously constitute an integral part of the Judeo-Christian heritage and play a role in the teaching of Christ Himself. He ought also to have realized that these juridical conceptions do not cover the entire ground, and that it was never intended that they should. And that is of course why so many metaphors, so many symbols, so many parables have been put before us. Each of these has something unique and precious to say: Christ did not speak vain words! And yet Teilhard takes it upon himself to dismiss the entire gamut of juridical

---

32. 1 Corinthians 13:12.
33. Romans 1:20.

images as a mere vestige of a so-called "neolithic symbolism"[34] which he declares to be outmoded.

In point of fact, authentic symbolism *never* becomes outmoded: for it is rooted in the nature of things. The worst that can happen is that in a superficial age—an era in which spiritual vision has become well nigh a thing of the past—there may be hardly anyone left to "read the icon." And when this comes to pass, the religious outlook—if it survives at all within the more educated strata of society—does in fact become impaired. As Vladimir Lossky explains: "A theology impoverished by that rationalism which recoils before these, the images of the Fathers, necessarily loses the cosmological perspective of Christ's work."[35] Now this may well be in the main the kind of theology to which Teilhard had in fact been exposed during his seminary years. But this does not alter the fact that when he objects—quite rightly!—to a Christianity which had all but lost "the cosmological perspective of Christ's work," the beam is actually in his own eye.

What also displeases him are the so-called miracles, beginning with the miracles of Christ. Teilhard is careful for the most part not to deny outright that such things have taken place, yet he is obviously at pains to minimize their importance. Repeatedly he informs us that miracles, though they may have played a certain role in the early development of Christianity—on account of their "propaganda value" as one might say—have all but lost their significance in the present scientific age. But here again he is overlooking something of the utmost importance: miracles are profoundly significant insofar as they render visible, as it were, the reality of the spiritual world. Diseases which Nature could only heal slowly or not at all are cured in a trice; future happenings which seem to hinge upon a host of imponderables—like the crowing of the cock in the Gospel narrative—are perceived in advance; and much else which is no less an affront to our scientific wisdom is rendered manifest. The crucial question, of course, is whether such things

---

34. See for instance CE, p. 202.
35. *Orthodox Theology* (Crestwood, NY: St. Vladimir's Seminary Press, 1978), p. 202.

have actually taken place.[36] But assuming that they have, it is simply absurd to perceive in these preternatural happenings a psychological device to prod people into accepting the Christian religion. Granting that the miracles in question may have that effect, what is of far greater moment is the fact that these eruptions of the miraculous reveal to us—more eloquently perhaps than any metaphysical argument—the stupendous inadequacy of our ordinary naturalistic notions, the very conceptions which Teilhard tends to absolutize and extol.

Finally, Teilhard is not especially pleased with our Christian saints. Even their charity offends him: "It is the fire of a love which is almost exclusively 'ascensional' in type, its most operative and most significant act being always represented in the form of a painful purification and a joyless detachment." And he adds: "For the neo-humanists we all are now, this soon produces an atmosphere which we find unbreathable, and *it must be changed*."[37] What apparently bothers Teilhard the most is that these fervent men and women of "ascensional" type are not especially concerned with what he deems to be human progress: they generally display little interest in technology and "socialization." Worst of all, there is reason to believe that they are wont to commune with God in ways distinctly beyond our ken. Let us admit it outright: the saints are a living reproach, not just to Teilhard de Chardin, but to all of us: to all who are not as heroic, not as self-sacrificing, not as pure, and not as intoxicated with the love of God. And if we happen *not* to be neo-humanists, we find in them—not an "unbreathable" atmosphere of gloom—but rather a shining example, a living inspiration, and a perennial source of strength and joy.

There is something else that needs also to be pointed out in this connection: the saints exemplify at times a spiritual knowledge which is literally unimaginable, a knowing of which our Nobel lau-

---

36. We obviously cannot enter at this point into a discussion of that issue. Suffice it to say that there has been no dearth of the miraculous even in this day and age, and that a vast number of *bona fide* miracles have in fact been documented; one need but study the life of St. Pio of Pietrelcina (familiarly known to thousands as Padre Pio) to be convinced of this.

37. CE, p. 217.

reates have not a clue. We must not think of this as something "mystical" in the popular sense: it is not a cognition of vague things in the sky. And what is perhaps most marvelous of all, while that cognition does no doubt penetrate into the invisible reaches of the subtle and spiritual realms, it does not on that account lose touch with the corporeal sphere. The *bona fide* contemplative is able, no less than we, to perceive the realities of this world; but he can do so by different means, and in a different manner. To quote from one of the Byzantine masters: "He beholds this multitude of things and all this perceptible world, not by perception, nor by thought, but by the power and grace proper to the God-like intellect, which makes distant things [appear] as if before their eyes, and in a manner beyond Nature presents things to come as if they were already there."[38] No use trying to picture to oneself what such knowledge might be like; as an Indian sage once put it: "You cannot pour four seers of milk into a three-seer pot."

The Christian life is the following of Christ: it is the Way of the Cross. And Teilhard agrees. But what does he have in mind? Here is what he says:

> If you ask the masters of the ascetical life what is the first, the most certain, the most sublime of mortifications, they will all give you the same answer: it is the work of interior development by which we tear ourselves away from ourselves, leave ourselves behind, emerge from ourselves. Every individual life, if lived loyally, is strewn with the outer shells discarded by our successive metamorphoses—and the entire universe leaves behind it a long series of states in which it might well have been pleased to linger with delight, but from which it has continually been torn away by the inexorable necessity to grow

---

38. St. Gregory Palamas, quoted by Archimandrite Vasileos in *Hymn of Entry* (Crestwood, NY: St. Vladimir's Seminary Press, 1984), p. 114.

greater. This ascent in a continual sloughing off of the old is indeed the Way of the Cross.[39]

Let us note, in the first place, that this account makes no reference, open or implied, to the Crucifixion: Teilhard speaks of the Way of the Cross as if Jesus of Nazareth had never lived. Unlike the authentic Christian "masters of the ascetical life," he seems to have totally forgotten that Christ "suffered under Pontius Pilate" and was crucified. It seems that his faith in the divinity of the "historic Christ" does not entail the slightest belief in the saving efficacy of the Sacrifice for the sake of which He was born.

But Teilhard is also forgetting something else: there is more than one way of "sloughing off" shells. Certainly the Christian ascetic is discarding something of himself; but this alone does not suffice to characterize the Christian trajectory. If it were true, as Teilhard implies, that a single law of metamorphosis is the universal rule of life, then every caterpillar would basically be doing just what the Christian ascetic does—which is of course precisely what Teilhard is driving at. As he tells us himself: "All that rather involved way of putting it is simply a way of expressing the most commonplace and frequently met experience of our lives—the painfulness of hard work."[40] Amazing! The Mystery of the Cross reduces ultimately to "the painfulness of hard work." Never mind the unbelievable impertinence: we arrive in the end, as always, at a complete banality.

This scandalous interpretation of the Crucifixion is of course entirely consonant with Teilhard's doctrine of "creative union" and the misbegotten notion that "every process of material growth in the universe is ultimately directed towards spirit, and every process of spiritual growth towards Christ."[41] Does he not know that the world is full of enterprises in which material growth is coupled with a blighting of the spirit? And is he ignorant of the fact that everywhere one finds individuals who, be it out of foolishness or greed, work destruction? And this man calls himself an empiricist!

---

39. SC, p. 69.
40. Ibid.
41. SC, p. 68.

On *a priori* grounds, as we have noted before, Teilhard has locked himself into a position which forces him to maintain that all vectors point in one and the same direction. In Teilhard's one-dimensional universe there is only one way to go: one destination, one Point Omega down the universal road. In such a constricted universe—such an incredibly "narrow world"—the Way of the Cross can no longer be conceived. Does not a cross, by its very nature, entail *two* dimensions: a vertical and a horizontal namely? Now this iconographic fact, simple and obvious though it be, has actually a profound metaphysical significance: it teaches us that the Christian path—which *is* the Way of the Cross—is indeed set at right angles to the plane of the world. And so, too, it leads *out* of the world, not simply at the end, but right from the start: to become a Christian through baptism is already to depart in some measure from the horizontal plane of this world.

Notwithstanding all the considerations brought forth by Teilhard de Chardin, there is then, after all, a difference between the Way of the Cross and the metamorphosis of insects! Nor is the cosmos at large destined to complexify itself into the Mystical Body. And it is above all a false and dangerous idea that the voice which presently beckons humanity to pursue the twin goals of technological progress and "socialization" is in truth the Voice of Christ.

Here and there, in some of his more intimate writings, Teilhard reveals himself in the posture of a mystic; and who can tell whether this singularly extraordinary person may not indeed have been gifted with some kind of preternatural sight. We have no reason, surely, to doubt his word when he intimates that experiences of a mystical nature did play a decisive role in the formation of his doctrine, and that he perceived in these revelations a direct confirmation of his most essential beliefs.

There is an interesting parallel in this regard between Teilhard de Chardin and Carl Jung, who likewise presented himself as a man of science and propounded his far-flung theories on purportedly scientific grounds—only to let it be known in the end that he was in truth a prophet. Jung also, moreover, billed his doctrine as a kind of grand synthesis between science and religion, and as Philip Rieff observes, "supplied a parody of Christianity."[42] In Jung's case the definitive statement on the subject was issued in his famous memoirs, *Memories, Dreams, Reflections*, dictated in the last days of his life. And Rieff might well be right when he notes with irreverent candor that this revelation had been carefully planned. "To avoid martyrdom," he conjectures, "Jung delayed announcing his full membership in the confraternity of prophets until after his death, by arranging a posthumous publication of his autobiography, which is at once his religious testament and his science, stated in terms of a personal confession."[43]

There is an analogy, too, between the Jungian *Memories* and Teilhard's most intimate essay, "The Heart of Matter," which belongs to the last period of his literary career. Here as well one can speak of a posthumous publication "which is at once his religious testament and his science, stated in terms of a personal confession."

To be sure, one must not press these parallels too far. The element of "personal confession" and the detailed glimpses into his mystical workshop to be found in Jung's autobiography exceed by far what Teilhard has to offer along such lines. And despite the indicated similarities it must not be forgotten that the two men represent very different intellectual and spiritual types. Never, for instance, could Jung have extolled such things as Baconian science and the unbridled proliferation of technology, let alone the formation of totalitarian states. And the Swiss psychiatrist—who referred to Darwinism as "a religion, or—even more, a creed which has absolutely no connection with reason"[44]—might have been less-than-complimentary in his diagnosis of the French prophet had he left us one. Teilhard, on the other hand, would have been profoundly offended by Jung's alchemical speculations, by his generally high regard for "primitive" societies and all manner of ancient lore, by his scathing critique of modernism in many of its manifesta-

42. *The Triumph of the Therapeutic* (NY: Harper & Row, 1968), p. 139.
43. Ibid.
44. *Modern Man in Search of a Soul* (NY: Harcourt Brace, 1933), p. 175.

tions, and by a number of other Jungian traits, which are obviously opposed to his own.[45]

Teilhard's mysticism, too, is of a different stamp. It is much less "pictorial" and far less intricate from a symbolist point of view than the visionary experiences recorded in Jung's autobiography. Teilhard seems to be concerned more with concepts than with symbols of an iconographic kind. We are told, for example, how *Cosmic Convergence* and *Christic Emergence* "made themselves felt in the very core of my being," and how "they reacted endlessly upon one another in a flash of extraordinary brilliance, releasing by their implosion a light so intense that it transfigured (or even 'transubstantiated') for me the very depths of the World."[46] Yet despite their seemingly amorphous character experiences of this kind were no doubt immensely significant in Teilhard's eyes: "How is it, then," he exclaims, "that as I look around me, still dazzled by what I have seen, I find that I am almost the only person of my kind, the only one to have *seen*?"[47] To be "the only one to have *seen*"—in this expressive phrase Teilhard has undeniably staked out his prophetic claims. "I cannot, when asked, quote a single writer," he goes on to say, "a single work, that gives a clearly expressed description of the wonderful 'Diaphany' that has transfigured everything for me." In him alone, we are told two pages later, have "love of God and faith in the world" come together so as to fuse spontaneously. And Teilhard predicts that what has thus far happened only in himself will eventually take place on a grand scale: *"Sooner or later there will be a chain reaction"* he declares (in italics). "This is one more proof"—so reads the concluding line of "The Christic," completed one month before his death—"This is one more proof that Truth has to appear only once, in one single mind, for it to be impossible for anything ever to prevent it from spreading universally and setting everything ablaze."

In the face of such impassioned declamations one can hardly

---

45. For an analysis of Jung's doctrine and *modus operandi* I refer to my chapter on Jung in *Cosmos and Transcendence* (Tacoma, WA: Sophia Perennis/Angelico Press, 2008).
46. HM, p. 83.
47. Ibid., p. 100.

doubt the sincerity and indomitable force of Teilhard's prophetic convictions. Here we come face to face, not with a scientistic pretender, one who "cheats with words," but with a mystic of sorts, a soul afire. What then—the question can scarcely be avoided any longer—what could be the source of these compelling experiences, these veritable seizures? This is ever the crucial issue when it comes to mysticism.

Now, it is certainly not my intention to propose a definitive answer to this delicate and somewhat uncomfortable question. Suffice it to point out certain signs which seem to be significant. In particular, I would draw attention to one of Teilhard's early compositions, a piece entitled "The Spiritual Power of Matter" (dated August 8, 1919), which seems to be a dramatized account of a mystical experience through which Teilhard had recently passed. And significantly enough, Teilhard himself has appended this piece to "The Heart of Matter"—his "Confessions"—to "express," as he says, "more successfully than I could today the heady emotion I experienced at that time from my contact with Matter."[48] Let us see what Teilhard has to say in this illuminating "fantasy."

"The man was walking in the desert, followed by his companion, when the Thing swooped down on him": so it begins. We need not concern ourselves with all the dramatic particulars—what interests us is the impact of this strange encounter upon "the man." It has been impressively described:

> Then, suddenly, a breath of scorching air passed across his forehead, broke through the barrier of his closed eyelids, and penetrated his soul. The man felt he was ceasing to be merely himself; an irresistible rapture took possession of him as though all the sap of all living things, flowing at one and the same moment into the too narrow confines of his heart, was mightily refashioning the enfeebled fibres of his being.[49]

A striking passage, to be sure, which could easily have come from the pen of a *bona fide* Christian mystic. But let us go on: "And at the

48. Ibid., p. 61.
49. Ibid., p. 68.

same time the anguish of some superhuman peril oppressed him, a confused feeling that the force which had swept down upon him was equivocal, turbid, the combined essence of all evil and all goodness." Could this, too, have come from the pen of a Christian mystic? A presence which oppresses the soul and gives rise to confusion, a force that is "equivocal, turbid, the combined essence of all evil and all goodness"—could this be an Angel of Light? "You called me: here I am," says "the Thing"; "grown weary of abstractions, of attenuations, of the wordiness of social life, you wanted to pit yourself against Reality entire and untamed," the young seer is told in this distinctly Faustian scene. The spirit himself, moreover, disclaims his own holiness: "I was waiting for you in order to be made holy," he declares. "And now I am established on you for life, or for death.... He who has once seen me can never forget me: he must either damn himself with me or save me with himself." To which the seer replies: "O you who are divine and mighty, what is your name? Speak." Is it not strange that Teilhard should address as "divine" a spirit that is not holy, and stands himself in danger of being damned?

Such are the "signs" Teilhard the mystic has left behind: meager if you will, but by no means insignificant. In the final count it may be the laconic words of Hermes to Prometheus that hold the key: "It appears you are stricken with no small madness...."[50]

---

50. Aeschylus, *Prometheus Bound*, 977.

APPENDIX
THE RIDDLE OF GENESIS 2:4–5

I WISH NOW to show that, after presenting the *hexaemeron* account of Creation in Chapter 1, Genesis alludes—ever so discreetly—to the *omnia simul* doctrine, which may be termed "esoteric" in relation to the former: what indeed could be more so than a cosmogony exemplifying, if you will, the "perspective" of God Himself! Here, then, is the decisive text from Genesis 2:4–5:

> These are the generations of the heaven and the earth, when they were created in the day the Lord God made the heaven and the earth, and every plant of the field before it sprung up in the earth, and every herb of the ground before it grew.

According to the exegesis we are about to give, this is spoken from a "supra-cosmic" point of view: from "the standpoint of eternity" there are no longer *six* days, but only *one*.

It needs first of all to be pointed out that the terms "plant" and "herb" in Genesis 2:5 admit of a symbolic interpretation which proves to be crucial: the apparent exclusion of animals should itself alert us to the fact that more than the literal sense stands at issue. Why, then, this seeming restriction to plants? The reason appears to be that a plant (a *virgulum* or *herba*, to use the Vulgate terms), in marked contrast to animals, exists as it were in two domains: above ground, namely, and beneath the earth; and whereas existence "above ground" is evidently suggestive of the phenomenal sphere, the domain of visible manifestation, the latter must then refer to a transcendent realm of causes, an invisible domain wherein the seeds of living beings subsist, and wherein also they incubate and begin to sprout. What confronts us in these verses proves thus to be an "icon" of the perennial ontology: through the figure of a natural symbolism, Genesis 2:5 is actually speaking of metaphysical truths.

One should note that in this interpretation, "seeds" correspond precisely to what Patristic tradition terms *rationes seminales*, which constitute the essential reality of the creature as it emerges directly from the primordial creative Act "before it sprung up in the earth." Let me emphasize that these are not *physical* seeds, not physical entities at all: they are situated "below ground," after all, which is to say that they belong to a prior ontological plane. One must bear in mind, moreover, that whereas the physical or corporeal domain is subject to the spatio-temporal condition, "below ground" there is neither spatial separation nor temporal sequence in a literal sense. Here, beyond the confines of the physical or corporeal world, one enters the primordial creation—the corona, if you will, of God's creative Act—where everything is still "fused without confusion" to put it in Meister Eckhart's words. Such, in brief, is the metaphysical panorama revealed in the second chapter of *Genesis* according to this interpretation.

To argue the case, it is first of all needful to consider the aforesaid quotation in the context of the first three chapters, beginning with the remainder of Genesis 2:5, which we have thus far left out of account. The full verse reads as follows:

> And every plant of the field before it sprung up in the earth, and every herb of the ground before it grew: for the Lord God had not rained upon the earth; and there was not a man to till the earth.

Two events must consequently take place before plants can "spring up in the earth" and herbs can "grow," the first being that it must "rain upon the earth," and the second that a man must "till the earth." And both did come to pass, as we learn from the Genesis text that follows. The first is related in the very next verse, which reads: "But a spring rose out of the earth, watering all the surface of the earth." Should anyone object on the grounds that this "spring" is not rain, I would point out that the initial "but" is there to connect verse 2:6 with 2:5. We are thus told that the first requisite did come to pass, that "all the surface of the earth" was in fact "watered." Admittedly this "watering" did not come about through "rain" in our sense: how could it, seeing that *our* world did not yet exist! Of the two "events," then, which must come to pass before plants can

APPENDIX THE RIDDLE OF GENESIS 2:4–5   251

"spring up in the earth" and herbs can "grow," the first—which comprises God's work—came to pass immediately.

This brings us to the second requirement. Though not instantly satisfied like the first, it also is soon realized: for as we read in verse 3:23, ere long Adam was expelled from the garden of Eden for his transgression. In fact, the Lord God sends him out "*to till the earth from which he was taken.*" Here, then, is the missing piece of the puzzle. We learn now that the "springing up of plants" and the "growth of herbs" referred to in verse 2:5 takes place *after* the Fall: *not* in the primordial creation, symbolized by the Garden of Eden, but in the post-Edenic realm, precisely, which is none other than *our* world. It turns out that what verse 2:5 speaks of symbolically as "above ground" refers in fact to this spatio-temporal universe, which the metaphysically uninstructed take to be all.

The question remains why there should be a double reference to "plant" and "herb" (*virgulum* and *herba*, respectively, in the Vulgate): what could be the meaning of this duplication? In light of the elucidation already achieved, the explanation is not far to seek: for inasmuch as the verse refers to living creatures at large—to animals as well as plants, namely, both of which, for the stated reason, it designates "plants"—one of the two so-called plants must stand for animals: is it the *virgulum*, then, or is it *herba*? And as it happens, the text itself tells us so quite clearly by the adjectives *agri* and *regionis* it employs to characterize the two: for whereas an "agricultural" plant is evidently an authentic plant like wheat or barley, the adjective *regionis*—the signification of which is primarily geometrical, as in *recta regione*, meaning straight line—has evidently a very different ring. Basically it refers to a geometric limit or boundary of some kind. An *herba regionis* must consequently be a plant restricted to some locus, as in fact happens typically in the case of animals inasmuch as they are situated literally "above ground." An animal—from a fish to a terrestrial mammal—is thus an *herba regionis*. It is to be noted, moreover, that this interpretation is immediately confirmed by the words "before it grew" which replace the phrase "before it sprung up" previously applied to the *virgulum*: whereas the latter "springs up in the earth" before emerging from its subterranean ambience, the *herba* needs merely to "grow," which is to say that it was situated "above ground" from the start. It is to be noted that this fundamental distinction between plants and animals is moreover reflected in the basic fact that plants enjoy ecological priority: one might say that inasmuch as the source of life resides indeed "below ground," animals at large are connected ecologically to that source precisely by way of plants.

Based upon the Vulgate text, at least, the stipulated association of *virgulum* and *herba* with plants and animals, respectively, appears thus to be well founded. What all but settles the question however, in my view, is the fact that this exegesis accords both with the *omnia simul* teaching of Christianity and with the major metaphysical traditions of mankind.

One final remark: the interpretation of Genesis 2:4–5 which we have offered suffices to expose the fundamental fallacy of evolutionism, *be it Darwinist or theistic*. Either doctrine may henceforth be likened to a botany which knows nothing of seeds and roots.

PIERRE TEILHARD DE CHARDIN
BIOGRAPHICAL FACTS

PIERRE TEILHARD DE CHARDIN was born on May 1, 1881, at Sarcenat, near Orcines, Puy-de-Dôme, in south-central France. His father was a gentleman farmer with an interest in geology, and his mother a descendant of Voltaire. Teilhard was educated at the Jesuit College at Mongré, and joined the Society of Jesus in 1899. He studied philosophy and continued his seminary education from 1901 to 1905 at the Jesuit house on the Isle of Jersey. This was followed by a three-year sojourn in Cairo, Egypt, where he taught physics and chemistry at a Jesuit school and developed his paleontological interests. Teilhard returned to England in 1908, studied theology at Hastings, and was ordained in 1911. He subsequently returned to Paris and devoted himself to the study of paleontology at the Museum of Paris under the direction of Marcellin Boule, a noted authority of the time. These studies were interrupted by the outbreak of World War I. Refusing to serve as an army chaplain, Teilhard joined the French forces as a stretcher bearer. In recognition of his bravery he was decorated and received into the Legion of Honor.

Teilhard continued his paleontological studies after the war, and took a doctorate in 1922 from the Sorbonne. For a brief period he taught geology at the Catholic Institute in Paris. His less-than-orthodox theological opinions, however, especially with reference to Original Sin, led to the termination of this employment and his *de facto* exile to China. In 1923 Teilhard arrived thus in Tientsin, where he took up research as an assistant to the Jesuit paleontologist Émile Licent. He subsequently collaborated in the excavations at Choukoutien which led to the discovery of the so-called *Sinanthropus* ("Peking Man"), supposedly a "missing link" in the evolution of man from subhuman ancestors. And although this presumed discovery was later challenged (if not indeed disproved), it caused

Teilhard to become widely looked upon as a paleontologist of note. His reputation was further enhanced in 1931 when, together with the well-known Abbé Henri Breuil, he supposedly established that *Sinanthropus* had known the use of fire and primitive tools.

During World War II Teilhard continued his activities in Peking, where on account of the Japanese occupation he lived in virtual captivity. He returned to France in 1946 and tried unsuccessfully to gain permission from the Church for the publication of his philosophical writings and to secure a teaching post at the Collège de France. Unpublished copies of his numerous writings were widely circulated, however, and commenced to arouse great admiration and enthusiasm in Catholic circles, beginning with high-placed members of the Jesuit Order. In 1952 Teilhard accepted a position with the Wenner Gren Foundation for Anthropological Research in New York. He died in that city on Easter Sunday, 1955.

ACKNOWLEDGMENTS

THE AUTHOR AND PUBLISHER gratefully acknowledge permission to reprint excerpts from the following material:

Activation of Energy by Pierre Teilhard de Chardin, copyright © 1963 by Editions du Seuil; English translation copyright © 1970 by William Collins Sons & Co. Ltd., London. Reprinted by permission of Harcourt Brace Jovanovich, Inc.

Christianity and Evolution by Pierre Teilhard de Chardin, copyright © 1969 by Editions du Seuil; English translation copyright © 1971 by William Collins Sons & Co. Ltd., and Harcourt Brace Jovanovich, Inc. Reprinted by permission of Harcourt Brace Jovanovich, Inc.

Human Energy by Pierre Teilhard de Chardin, copyright © 1962 by Editions du Seuil; English translation copyright © 1969 by William Collins Sons & Co. Ltd., London. Reprinted by permission of Harcourt Brace Jovanovich, Inc.

The Heart of Matter by Pierre Teilhard de Chardin, copyright © 1976 by Editions du Seuil; English translation copyright © 1978 by William Collins Sons & Co. Ltd., and Harcourt Brace Jovanovich, Inc. Reprinted by permission of Harcourt Brace Jovanovich, Inc.

The Future of Man by Pierre Teilhard de Chardin. Translated by Norman Denny. Copyright © 1964 in English translation by William Collins Sons & Co. Ltd., and Harper & Row, Publishers, Inc. Reprinted by permission of Harper & Row, Publishers, Inc.

The Phenomenon of Man by Pierre Teilhard de Chardin. Copyright © 1959 in the English translation by William Collins Sons & Co. Ltd., and Harper & Row, Publishers, Inc. Reprinted by permission of Harper & Row, Publishers, Inc.

LIST OF ABBREVIATIONS
FOR THE WRITINGS OF
TEILHARD DE CHARDIN

AE: *Activation of Energy*
(NY: Harcourt Brace Jovanovich, 1970)

CE: *Christianity and Evolution*
(NY: Harcourt Brace Jovanovich, 1971)

DM: *The Divine Milieu*
(NY: Harper & Row, 1968)

FM: *The Future of Mankind*
(NY: Harper & Row, 1964)

HE: *Human Energy*
(NY: Harcourt Brace Jovanovich, 1969)

HM: *The Heart of Matter*
(NY: Harcourt Brace Jovanovich, 1979)

MN: *Man's Place in Nature*
(NY: Harper & Row, 1966)

PM: *The Phenomenon of Man*
(NY: Harper & Row, 1965)

SC: *Science and Christ*
(London: Collins, 1968)

INDEX OF NAMES

Abraham 146
Adam 148–153, 159, 161, 251
Aeschylus 247
Aquinas, St. Thomas 51, 66, 112, 160
Aristotle 49, 67, 92
Arius 9
Augustine, St. 20, 21, 24, 25, 50, 118, 128, 142, 159, 181, 182, 212

Bacon, Francis 181
Balthasar, Hans Urs von 138
Basil the Great, St. 86
Behe, Michael 4, 16
Benedict XVI, Pope: *see* Cardinal Ratzinger
Bertalanffy, Ludwig von 29
Bérulle, Pierre de 122, 123
Bohm, David 127
Bohr, Niels 35
Boule, Marcellin 253
Bounoure, Louis 27, 191
Breuil, Abbé Henri 254
Burckhart, Titus 28, 183

Capra, Fritjof 127
Coomaraswamy, Ananda 25, 112, 183
Copernicus, Nicolaus 148

Danielou, Jean Cardinal 73
Dante Alighieri 86, 113
Darwin, Charles 3, 7, 16, 17, 19, 26, 27, 31, 148, 234

Democritus 92
Denton, Michael 16, 18
Descartes, René 39, 44, 58, 92, 171, 192
Dewar, Douglas 16
Dionysius the Areopagite 131, 141
Drexel, Albert 226

Eckhart, Meister 20, 129, 159, 250
Einstein, Albert 7, 44, 197
Ellul, Jacques 177
Erasmus, Desiderius 168
Eve 148, 153, 159, 161

Florovsky, George 158, 229
Freud, Sigmund 68, 123

Galileo 92, 171
Gillespie, Charles 181
Gray, James 30
Gregory of Nyssa, St. 161

Haeckel, Ernst 17, 27
Hasak, Maximilian 176
Hawking, Stephen 7, 197
Heraclitus 82, 92, 126
Herbert, George
Hermes 247
Hildebrand, Dietrich von 215, 223, 236
Hildegard, St. 170, 171
Hitler, Adolf 198, 200
Huxley, Julian 69, 72

Jaki, Stanley 30
Jerome, St. 9
John the Apostle, St. 135
John the Baptist, St. 204
John XXIII, Pope 226
Johnson, Phillip 16
Jung, Carl 68, 123, 243–245

Kepler, Johannes 182
Koestler, Arthur 29

Lamarck, Chevalier de 26
Lauriers, Guerard de 115
Leucippus 92
Lewontin, Richard 5
Lings, Martin 183
Lossky, Vladimir 158, 161, 239
Lubac, Henri Cardinal de 78, 101, 115, 116, 191

Mao Tse Tung 200
Marx, Karl 12
Maximus, St. 44
Medawar, Peter 11, 69, 189, 235
Mendel, Gregor 3
Mersch, Émile 144
Mignot, Jean 183
Milton, Richard 16
Moses 57, 80, 118
Mozart, Wolfgang 41

Nasr, Seyyed Hossein 163, 167, 169, 170, 176, 183

Palamas, Gregory 241
Parmenides 92
Patterson, Colin 16
Paul, St. 47, 48, 94, 112, 130, 143, 159, 163, 170, 173, 197, 205, 216, 227, 228, 238
Penfield, Wilder 40, 41, 57
Pio of Pietrelcina, St. 240
Plato 57, 92, 181, 182
Prometheus 191, 207–209, 211, 247
Ptolemy, Claudius 148

Ratzinger, Joseph Cardinal 226–227
Richard of St. Victor 157
Rieff, Philip 244
Rolt, C. E. 141
Rostand, Jean 28, 29
Roszak, Theodore 174, 175, 177, 181
Rousseau, Jean Jacques 168

Schuon, Frithjof 164
Sherrington, Charles 41, 58
Simpson, George Gaylord 26, 109, 115, 124
Smith, Huston 29, 46, 168, 187, 223
Stalin, Joseph 200

Taylor, Sherwood 170

Vasileios, Archimandrite 219
Vernet, Maurice 58
Voegelin, Eric 196
Voltaire, Françoise 253

Whitehead, Alfred North 238

Zanta, Leontine 224